Emergency One Fire Apparatus

1974-2007

Kent Parrish

Iconografix

Iconografix
PO Box 446
Hudson, Wisconsin 54016 USA

Library of Congress Control Number: 2008933687

ISBN-13: 978-1-58388-225-2
ISBN-10: 1-58388-225-1

08 09 10 11 12 13 6 5 4 3 2 1

Printed in China

Cover and book design by Dan Perry

Copyedited by Andy Lindberg

On the cover: from left to right, top to bottom.

1. Folsom, California; 1985 Hurricane; 1500 gpm pump; 500-gallon water tank; S/O 4000. *Chuck Madderom*
2. Fort Knox, Kentucky; 2004 Titan HPR 4x4; 1500 gpm pump; 1500-gallon water tank; 200-gallon foam cell; 500-pound dry chemical system; S/O 127530. *Kent Parrish*
3. Kirkwood, Missouri; 1993 Cyclone; 1500 gpm pump; 450-gallon water tank; 50-gallon foam cell; 75-foot aerial ladder; S/O 12576. *Dennis J. Maag*
4. Willow Springs, Illinois; 2001 Cyclone II; 1500 gpm pump; 300-gallon water tank; 95-foot aerial platform; S/O 122412. *Garry Kadzielawski*
5. Washington D.C.; 1988 Hush hazardous materials unit; S/O 5464. *Joel Woods*
6. Rincon Del Diablo, California; 1979 GMC 7000 4x4 20084 midi-pumper; 250 gpm PTO pump; 500-gallon water tank; S/O 976. *Chuck Madderom*

TABLE OF CONTENTS

DEDICATION

This book is dedicated to my bride and best friend Kimberlie, who has taken a kind interest in my passion.

ACKNOWLEDGEMENTS

The completion of this book would not have been possible without the assistance and diligent photography of many friends and fellow fire apparatus buffs. Credit is given with each image that appears in this book. Additional appreciation is extended to Bill Friedrich who loaned the use of his collection of vintage manufacturer literature. Gratitude must also be conveyed to Ron Arnow, Tommy Barnett, and Tom Shand who provided invaluable information.

BIBLIOGRAPHY

Pumpers – Workhorse Fire Engines, by Larry Shapiro, MBI Publishing Company, 1999
Aerial Fire Trucks, by Larry Shapiro, MBI Publishing Company, 2002
My Friend the Genius, by Tom Hill, 2004-2006

FOREWORD

Robert Wormser, the founder of Emergency One, once received an antique fire engine as a gift from an investor. He had the rig restored and displayed it briefly at the plant. Much to the dismay of classic buffs, he then put the apparatus in storage, saying he "didn't like looking back, only forward with new innovations."

INTRODUCTION

As the 1970s began, the American fire apparatus industry was dominated by well-established manufacturers whose tried and true methods consisted of building classics out of steel. In 1974, a new player arose out of sunny Florida. This upstart and its use of modular aluminum bodies revolutionized how fire apparatus was built. By 1984, its Hurricane custom chassis, dubbed after the disastrous weather phenomena that frequent the Florida coasts, was an ironic descriptor of how Emergency One had blown through the industry and left the unprepared fire apparatus legends in its aftermath.

Robert S. Wormser was an accomplished welder, fabricator, and metal artisan. An inventor with over 100 patents to his name, he was also a clever promoter, motivator, and negotiator who had established himself in the commercial park and playground equipment industry. Wormser retired from the cold Michigan winters to the year round sun and fun of Ocala, Florida, in 1972.

Growing impulsive two years into retirement, Wormser set out to build an ambulance "because he'd never done so." He built a body out of aluminum in a garage in his backyard, placed it on a conventional van chassis, and sold it locally. Surprisingly there was a demand for more units. Wormser then incorporated a new company called Emergency One, which he named in part after a popular and influential television series called *Emergency!* featuring Station 51 of the Los Angeles County Fire Department.

After Wormser recruited a handful of former colleagues down to Ocala, Emergency One quickly grew to two Quonset huts and a small office building. However, the company soon found the ambulance market a purchasing nightmare because of the wide range of medical equipment specified by each customer to accompany its unit and Wormser decided to try his hand at building a fire truck. He designed an extruded aluminum body and constructed a mini-pumper on a pick-up chassis in his garage. Wormser then met with his crew and announced they were going to abandon ambulances and start building fire trucks, which would be "more fun." Emergency One's new logo would become the familiar warning beacon next to the company name, all within a large oval.

Emergency One started building fire trucks completely against the grain of the industry standard and in unprecedented fashion. The first units were quick attack mini and midi-pumpers on light- and medium-duty commercial chassis because they were readily available and serviceable. By using aluminum, Robert Wormser felt they could build a lighter and corrosion resistant fire truck that could carry more equipment and fare much better in areas with salty air and inclement weather. This selling point proved to be fact and beneficial when competitors later saw their apparatus fall prey to early rust as a result of bad steel used during the 1970s. It was also advantageous that Emergency One's birth came during a period when the fire service trend of using quick attack apparatus was in full swing.

Emergency One's fire apparatus bodies were formed of aluminum extrusions and welded panels. They were of modular construction meaning they were built independently of the chassis. This method resulted in a much faster build time. Employing this schedule and keeping a large inventory of chassis and pumps on hand, Emergency One could deliver complete apparatus in as little as 30 to 60 days instead of the industry standard of six months or more.

Robert Wormser grew his fledgling company with bold tactics. One early story tells of a local dealer coming by the shop one day to collect payment or repossess a group of pickups that had been delivered. Knowing this day could come, Wormser had foiled the dealer by quickly chopping off the rears of the pickups in preparation to become mini-pumpers. The dealer left empty handed that day, but eventually received payment as the units were completed and sold.

Emergency One quickly became legendary for its bold advertising and marketing gimmicks. This included the unprecedented giveaways of fire trucks at trade shows and turning all of the drawing entrants into sales leads. Another subliminal trick was flying West Coast fire chiefs to the plant and allowing them to deliver brand new Emergency One fire trucks to their new departments. Additional publicity was gained by loaning fire trucks loaded with Emergency One emblems to production studios for use in television shows and films. Emergency One was also an early pioneer in color advertisements and the mass mailing of brochures.

Robert Wormser also had a knack for showing up his vendors. Unhappy with the time lag in obtaining Hale fire pumps, Emergency One began to slowly increase its orders. After stocking a substantial inventory, Wormser then boasted in an ad that Emergency One had more pumps on hand than Hale itself! Another story tells of Wormser's disgust over warranty issues with the aftermarket retrofitting of front driving axles on GMC truck chassis to create 4x4 apparatus. When he requested that GMC supply him with factory direct 4x4 truck chassis and full warranties, he was told they would require an order for at least 50 units. So, Wormser ordered 50 units and had them delivered to his dealer. He then boasted Emergency One was the only company in the world with factory direct 4x4 GMC truck chassis!

Emergency One found an aptitude for financing its early growth. One boost came in the form of Nigeria taking bids on 40 armored riot vehicles equipped with high-pressure pumps and water tanks. Emergency One placed a ridiculously high bid which the Nigerians accepted! The units were successfully completed and accepted. Not only did this order reap a huge profit, it opened the door wide open to a foreign market, which Emergency One quickly took advantage of.

After just four years of business, Emergency One was bursting at the seams and moved to a brand new 82,000 square foot plant on a ten-acre site adjacent to Interstate 75 in Ocala, complete with baseball diamond and picnic area. Robert Wormser acted as the contractor during its construction and each employee contributed in some fashion. By all accounts, Emergency One was truly an enjoyable place to work and Wormser received the most out of his employees by treating them as family and placing himself at the bottom of the company hierarchy. The company made sure to display multicolored rows of brand new fire trucks with Federal Twinsonic lightbars in view of Interstate 75 awaiting delivery and glistening in the Florida sun

In order to finance future growth, Emergency One realized the need for a large investor and sold itself to the Federal Signal Corporation in 1979. This emergency equipment conglomerate could consistently bankroll the development needed to continue its thrust which culminated with Emergency One completely taking over the fire apparatus industry within just a decade of its founding.

In 1985, a spin-off company named Federal Motors began operations in a building at the Ocala plant in 1985 with Robert Wormser at the helm. It was tasked with producing custom fire apparatus chassis for Emergency One while branching out into specialized chassis for the industrial, municipal, and transportation markets. Around this time, the oval Emergency One emblem was updated to simply state "E-One" and by 1988 the new fire apparatus giant was officially known as such while touting itself as "America's Fire Truck."

Robert Wormser's unique influence diminished as corporate politics began to throttle E-One. Throughout the 1990s, Federal Signal feasted on major acquisitions that expanded its product line and opened new markets. The first was Superior Emergency Vehicles of Red Deer, Alberta, Canada, in 1992 which became E-One Canada in 2003. In 1996, Federal Signal purchased the Finnish-built Bronto Sky Lift aerial line. In 1998, Federal Signal acquired Saulsbury Fire Apparatus of Tully, New York. Saulsbury was noted for its high quality stainless steel fire apparatus bodies. Combined with E-One, these subsidiaries formed Federal Signal's Fire Rescue Group (FRG). Superior and Saulsbury continued to build apparatus under its own brand while also completing apparatus on E-One custom chassis. The Canadian plant would also build a line of quick attack units, tankers, and light rescues bearing the E-One name. Saulsbury would build bodies for the E-One Bronto aerials and specialty rigs. The New York plant was also base for certain E-One lines that would not be built in Ocala. Federal Signal closed the Saulsbury plant in 2004 and the Superior plant in 2006, plunging two great names into history, and consolidating operations in Ocala.

To date, E-One has produced over 25,000 emergency vehicles for agencies all over the world. However, over the last several years, E-One has been suffering from a revolving door of management and quality control issues resulting in a slip in market share and consistent earnings loss. Despite a rebound that has increased sales and improved quality, as of May 2008, its parent company, Federal Signal, has put E-One on the selling block. With its future relatively uncertain, hope remains that an invigorated buyer will keep E-One in the upper echelon of the fire apparatus industry.

Too young to label a legend, but no longer an upstart, E-One has spanned a life fruitful enough to compile a brilliant photographic composition of the rigs that made it a true pioneer in the modern fire apparatus genre. The late Robert Wormser once said it best – fire trucks are "fun," a statement from a man who, after all, retired from playground equipment!

Chapter 1

While the fate of the first couple of fire trucks built by Emergency One during 1974 is not known, it is certain this unit on 1975 Ford F-350 chassis for Brea, California, was among the earliest deliveries. Emergency One's early models were generally designated according to a combination of the gross vehicle weight of the chassis and compartment demographics. The 10060 and 10084 mini-pumpers came standard with 250 gpm PTO pumps and 260-gallon water tanks. The 10060 had a compartment above the wheel well and one behind it. The 10084 featured a third compartment between the pump panel and rear wheel. *Chuck Madderom*

QUICK ATTACK UNITS COMMERCIAL PUMPERS & TANKERS

During the infancy of Emergency One, records were not officially kept. It is recalled by past and present employees that the first couple of fire trucks built by Emergency One were mini-pumpers delivered to volunteer fire departments, which no longer exist due to consolidation, in Volusia County, Florida.

Those first units were a representation of the early fire apparatus built by Emergency One and the foundation on which the company grew – quick attack units consisting of mini-pumpers built on pickup-truck chassis and midi-pumpers built on medium-duty conventional and tilt-cab commercial chassis. As these units proved successful, Emergency One naturally transitioned into full size Class A pumpers and tankers with Hale being the exclusive supplier of fire pumps from the beginning into the late 1980s. To date, Hale is still the majority pump supplier and Emergency One has built apparatus on practically every chassis commercially available in lines from the Sentry to Protector and under economy programs from the Town & Country to the Tradition line of today.

Michiana Shores, Indiana, received this 10060 mini-pumper on a 1975 Ford F-350 chassis with standard 250 gpm PTO pump and 260-gallon water tank. Early Emergency One fire apparatus featured crude bodywork with exposed aluminum welds and poorly sealed compartments. Removable stainless steel pump panels on the driver's side partially exposed the pump cavity. Emergency One was also an early catalyst for non-traditional paint schemes, from white to green. *Garry Kadzielawski*

Although Florida was certainly its natural haven, Emergency One found early success in California on the premise of its aluminum bodies faring better in salty sea air. Oxnard received this unit in 1975 on a 1974 GMC chassis. The 15060 and 15084 midi-pumpers came standard with 250 gpm PTO pumps and 300-gallon water tanks. The 15060 had compartments above the wheel well and behind it. The 15084 featured a vertical compartment between the pump panel and rear wheel. A variation was the 15084TC, on a Ford C-600 tilt-cab chassis. *Chuck Madderom*

Concord, Kentucky, purchased this 15060 midi-pumper with a 250 gpm PTO pump and 300-gallon water tank. A trait and telltale characteristic of Emergency One bodywork would become the diamond plating around the exterior of the rear wheel wells. Early models featured a more obtrusive, boxy design that covered the entire fender area. *Byron Brown*

This 15060 midi-pumper was delivered to Palm Bay, Florida, on a 1975 Chevrolet chassis with a 250 gpm PTO pump and 300-gallon water tank. The 10084 and 15084 designations would be phased out in favor of the standard of 10060 and 15060 for the mini- and midi-pumpers. *Glenn Vincent*

South Trail, Florida, received this 10060 mini-pumper on a 1975 Dodge 4x4 chassis with 250 gpm PTO pump and 260-gallon water tank. By late 1975, Emergency One had begun to refine its bodywork. Enhanced pump panels fully closed the pump cavity and became integral with the body. The body sat on a sturdier extruded aluminum frame and the characteristic diamond plate at the rear fender had become more elegant and flowed more in line with the wheel well. *Glenn Vincent*

With early success in quick attack units, it was only natural for Emergency One to segue into Class A pumpers. The stubby Emergency One 18060 Hellcat featured a 750 gpm front-mount pump and 500-gallon water tank. The Basalt and Rural Fire Protection District in Colorado acquired this unit on a 1976 GMC 4x4 chassis. Early Emergency One apparatus were not identified until a numerical sequence of shop numbers (S/O) was adopted with a starting point at number 100 sometime in late 1975. The rigs themselves were not consistently physically identified with these numbers until sometime late in 1976. This unit was assigned S/O 265. *Dennis J. Maag*

The Emergency One Guardsman Class A pumper was available on 25,080 to 27,000 GVW commercial chassis with pump capacity from 500 to 1000 gpm and water tanks up to 1000 gallons. Bagdad, Kentucky, received S/O 496 in 1977 on a 1976 GMC 6500 chassis with a front-mount 750 gpm pump and 750-gallon water tank. Because Emergency One stocked an inventory of commercial chassis, it was common to see apparatus built on a chassis a year or two older than the actual build date. *Kent Parrish*

The Sentry I was advertised as the industry's smallest Class A pumper. It featured a 500 to 1000 gpm midship pump and 500-gallon water tank on a 25,080 GVW commercial chassis with compact 149-inch wheelbase. Martinsville, Indiana, operated this Sentry I on a 1981 GMC 6500 chassis. S/O 1936 featured a 1000 gpm pump and 500-gallon water tank. *David Mattingly*

An upgrade was the Sentry III, which placed the model on a heavier commercial chassis with five-man canopy cab. Glenview, Illinois, operated this Sentry III on a 1981 Ford C-8000 chassis. S/O 2109 was equipped with a 1000 gpm pump and 500-gallon water tank. The Sentry III was also available with top-mount (TM) controls. *Chuck Madderom*

Next in the Emergency One Class A pumper line was the Protector I, available on a 25,080 GVW conventional commercial chassis. It was separated from the Guardsman by a transverse compartment between the cab and pump panel with cross-lay hose beds. Plainville, Indiana, operated this Protector I on a 1977 GMC chassis with 1000 gpm pump and 500-gallon water tank. *Frank Wegloski*

The Protector I was also available with top-mount (TM) controls. The Bayville Fire Company in Berkeley Township, New Jersey, operated this Protector I TM on a 1984 GMC 6500 chassis with 1000 gpm pump and 500-gallon water tank. *Scott Mattson*

Continuing up the Emergency One Class A pumper line was the Protector II, which was based on a 27,000 to 35,000 GVW commercial chassis with a two-door tilt-cab, the most prevalent choice being a Ford C-800 or C-8000. It featured a cross-lay hose bed with transverse compartment between pump and cab or top-mount control panel in lieu of. Pumps from 500 to 1000 gpm were available with water tanks up to 1000 gallons. Maryville, Illinois, operated S/O 380 on a 1980 Ford C-8000 chassis with a 1000 gpm pump and 750-gallon water tank. *Dennis J. Maag*

The Protector III was the big boy of the Emergency One commercial pumper line. It was based on a 27,000 to 35,000 GVW chassis with five-man canopy cab. The model featured pumps up to 1250 gpm with water tanks up to 1000 gallons. A cross-lay hose bed with transverse compartment was standard. This Protector III on a Ford C-8000 chassis was donated by Emergency One to the Fire Academy at the Pinellas County Vocational Institute in Florida. *Bill Friedrich collection*

By now, Emergency One offered nearly two dozen models and variations of commercial pumping apparatus. The 20102M midi-pumper was based on a 25,080 GVW conventional chassis and available in two- or four-wheel drive. It featured a mid-ship pump from 500 to 750 gpm and 300- to 500-gallon water tank. The Ridgeway Fire Company in Manchester Township, New Jersey, operated this 20102M on a 1981 GMC 4x4 chassis. *Frank Wegloski*

The Emergency One 20084 midi-pumper/rescue was the refined identification for the original 15084 design. In addition to a 250 to 750 gpm PTO pump and 500-gallon water tank, it featured ample compartment space for rescue equipment with a payload capacity of 5000 pounds. Jefferson Township at Otwell, Indiana, received this unit in 1978 on a Ford F model chassis. The choice of a 500 gpm mid-ship pump upgraded the model to a 20084M. *Frank Wegloski*

The 20084 was also available with four-wheel drive. Later models also saw the pump capacity increased to 350 gpm. Pawcatuck, Connecticut, operated this 20084 on a 1986 GMC 4x4 chassis with 350 gpm pump, 500-gallon water tank, high compartments, and dual booster reels. *Mark Redman*

A variation of the Emergency One midi-pumper/rescue was the 20084TC based on the Ford C-600 tilt-cab chassis. Palm Springs, California, received this unit in 1978 with a 250 gpm PTO pump and 500-gallon water tank. This view shows the full body compartment layout standard on the officer side of the Emergency One quick attack units. The 20108TC variation featured a transverse compartment between the cab and pump panel on the driver's side. *Chuck Madderom*

Desert Hills, Arizona, also received a 20084TC midi-pumper/rescue in 1978 on a Ford C-600 chassis with 250 gpm PTO pump and 500-gallon water tank. The compact features and maneuverability of this model proved to be popular with not only small departments, but big city departments such as Nashville and Jacksonville, which both ordered multiple units to run ahead of larger apparatus. *Chuck Madderom*

Robert Wormser aggressively marketed Emergency One by loaning units to production companies for use in television and film. This 15060 midi-pumper with a 250 gpm pump and 300-gallon tank on a 1977 GMC 4x4 chassis was used in the television pilot *Pine Canyon is Burning*. It was fictitiously lettered for the Los Angeles County Fire Department. *Chuck Madderom*

Commercial apparatus with larger pump capacities made economical alternatives for many big city/urban departments which were looking to replace large numbers of rapidly rusting steel bodied apparatus. Chicago, Illinois, received twenty of these 1250 gpm pumpers with Sentry III body styles and 500-gallon water tanks on Ford C-8000 chassis in 1980. They were assigned S/O numbers 1294–1303 and 1605–1614. *Garry Kadzielawski*

An early catalyst for Emergency One was its willingness to build unusual aluminum apparatus that its aging competitors would not engineer. This allowed them to corner the specialty market of industrial fire protection. The historic Climax Mine near Leadville, Colorado, was protected by this one-off unit built on a 1980 GMC 4x4 chassis. S/O 1880 featured a 1000 gpm pump, heated booster reel compartment, 1000-gallon water tank, and 40-gallon foam tank. *Dennis J. Maag*

Anne Arundel County, Maryland, operated this Protector II pumper with an unusual "pill box" style crew enclosure behind the International CO series tilt-cab in lieu of the transverse compartment and cross-lay hose bed. S/O 1890 was equipped with a 1250 gpm pump and 500-gallon water tank. *Chuck Madderom*

Emergency One was also among the first manufacturers to build pumpers equipped to transport medical patients. This "Ambulance Pumper" was one of two delivered to Winter Park, Florida, on 1981 Ford F model chassis with pill box style compartments fully equipped to load and transport patients on stretchers. They also featured 1000 gpm pumps with bumper mounted monitors and 500-gallon water tanks. *Garry Kadzielawski*

Emergency One offered tankers ranging from single rear axle units to massive tandem axle jobs with small PTO pumps to large mid-ship pumps. The West Routt Fire Protection District in Hayden, Colorado, is known for its fleet of unique Emergency One fire apparatus. This rig was constructed in 1981 on a 1977 Peterbilt over-the-road chassis with manual transmission. S/O 2015 was equipped with a 1500 gpm pump with top-mount controls, 2000-gallon water tank, and 120-gallon foam tank. *Dennis J. Maag*

Another marketing gimmick of Robert Wormser was a wildly popular and unprecedented annual drawing and giveaway of a free fire truck. Flintstone, Maryland, won the 1982 drawing at the IAFC show in Philadelphia for this 10060 mini-pumper on a Ford F-350 chassis with a 250 gpm pump and 260-gallon water tank. When Wormser handed over the keys to the winning department, there was no shortage of light-hearted "Fred Flintstone" and "Barney Rubble" jokes. *Shaun P. Ryan*

Donner Summit in the beautiful Sierra Nevada range of California was protected by this Guardsman pumper on a 1982 GMC 4x4 chassis. S/O 2865 was equipped with a 1000 gpm pump and 750-gallon water tank. After taking on GMC in regards to warranty issues on aftermarket systems, Robert Wormser would later boast that Emergency One was the only manufacturer to have factory-direct GMC 4x4 truck chassis. *Chuck Madderom*

This compact 15060 midi-pumper provided quick attack functions for bustling Seat Pleasant in Prince George's County, Maryland. "Big Minnie" was built on a 1982 GMC 4x4 chassis. It was equipped with a 250 gpm pump and 300-gallon water tank. *Joel Woods*

In addition to big tandem rear axle jobs, Emergency One also offered more compact tankers on single rear axles. Cedar Hammock, Florida, received this unit on a 1983 Ford C-8000 chassis with 250 gpm PTO pump and 1500-gallon water tank. *Leo Duliba*

This wildland brush rig for San Diego, California, was built off a modified design from the airport crash/fire/rescue (CFR) models with a sloped rear body for encroachment during off road operations. Built on a 1985 International S model 4x4 chassis, S/O 4079 was equipped with a 500 gpm pump with 600-gallon water tank. *Garry Kadzielawski*

In the mid 1980s, Emergency One offered an economical pumper called the U.S. One. It was available with a set option list to keep the cost down. The model could be differentiated by the lack of the characteristic diamond plate around the rear wheel well. South Windham, Connecticut, received S/O 5788 in 1985 on a GMC 7000 chassis with 1000 gpm pump and 750-gallon water tank. *Mark Redman*

By the mid 1980s, Emergency One had updated its builder plates to simply state "E-One." The Sandy Hook Fire Company in Newtown, Connecticut, received this big job on a 1986 Kenworth tandem axle chassis. S/O 5126 featured a 250 gpm PTO pump and 3000-gallon water tank. *Mark Redman*

The Jordan Fire Company of Waterford, Connecticut, operated this nice looking 15060 mini-attack pumper on a 1986 Ford F-350 4x4 chassis with 250 gpm pump and 250-gallon water tank. *Mark Redman*

By 1987, E-One Class A commercial pumpers were simply designated by one of two body styles in an effort to streamline the model lineup. The Protector XL featured a standard side-mount pump panel. Washington, D.C., received a total of twenty of these 1250 gpm Protector XL pumpers with pill-box style crew cabs and 500-gallon water tanks from 1986 to 1991. *Howard Meile*

The Protector XLT body style featured a top-mount pump panel. Dating back to the riot vehicles built for Nigeria in the 1970s, E-One had built a solid international presence. Milnerton, South Africa, purchased this Protector XLT on a 1989 Ford C-8000 chassis with a 3784 gpm (Imperial) pump, 2800-liter water tank, and 580-liter foam cell. *Shaun P. Ryan*

By 1987, E-One had generally phased out the numerical quick attack model numbers in favor of designating them by genre. This "midi-attack pumper" was designed with large flotation tires on a 1987 GMC 4x4 chassis for Carolla, North Carolina, to operate on the sandy beaches of the Outer Banks. S/O 6016 featured a 750 gpm pump and 500-gallon water tank. *Mike Sanders*

Baltic, Connecticut, received this ornate Protector XL pumper on a 1988 Ford C series chassis. S/O 6344 was equipped with a 1250 gpm pump and carried 1000 gallons of water. *Shaun P. Ryan*

Orange County, Florida, was naturally a large user of E-One apparatus. This big pumper-tanker was delivered in 1988 on an International S model chassis with four-door cab. S/O 6405 featured a 1250 gpm front-mount pump and 1250-gallon water tank. *Glenn Vincent*

This tanker was operated by Morgantown, Indiana. It was built on a 1989 Ford C series chassis. The apparatus was equipped with a 500 gpm pump, enclosed and heated booster reel compartment, and 1250-gallon water tank. E-One is reported to be the last fire apparatus manufacturer to build on the venerable Ford C series chassis, which was discontinued by 1991. *Frank Wegloski*

The Laurel Lake Fire Company of Commercial Township, New Jersey, received this big tandem axle tanker on a 1990 Ford L-9000 chassis canopy cab. It featured a 1250 gpm pump and 2500-gallon water tank. *Scott Mattson*

E-One delivered this unusual tanker with a very clean appearance to the Dodgington Fire Company of Newtown, Connecticut, on a 1991 White tandem axle chassis. S/O 8283 was equipped with a 450 gpm pump and 2500-gallon water tank with driver side jet dump. *Mark Redman*

In the early 1990s, E-One offered the Town & Country program apparatus with a set option list to provide a simple and economical apparatus. It was based on an International 4900 chassis with two- or four-door cab. Pumps with side- or top-mount controls were available up to 1250 gpm with tanks up to 1000 gallons. S/O 10521 with a 1250 gpm pump and 750-gallon tank went to Port Lavaca, Texas, in 1992. *Dennis J. Maag*

St. Marys, Georgia, received this mid-attack pumper/rescue in 1991 on an International 4700 chassis. S/O 10058 featured a 500 gpm pump, 300-gallon water tank, and 30-gallon foam cell. With the demise of the venerable Ford C series, International would become among the most prevalent choices for medium-duty commercial fire apparatus chassis. *Kent Parrish*

This tanker was delivered to Hamel, Illinois, in 1993 on a Ford L-8000 chassis. S/O 12295 was equipped with a big 1000 gpm front-mount pump and 1500-gallon water tank, all on a single rear axle chassis. *Dennis J. Maag*

This unique rig was built for Apache Junction, Arizona. S/O 14435 was built on a 1995 Freightliner 4x4 chassis with a sloped rear body for encroachment while off-road. The 1250 gpm unit carried 500 gallons of water and 20 gallons of foam. A remote controlled turret was mounted on the cab roof. *David Greenberg*

In addition to the International, the Freightliner Business Class was a prevalent commercial chassis choice in the fire apparatus industry. Pauls Valley, Oklahoma, received S/O 12218 in 1994 on a FL-80 chassis with a 1250 gpm pump and 750-gallon water tank. At this time, the commercial apparatus did not fall under specific model names or programs. *Steve Loftin*

This industrial pumper with international flair was delivered to the power plant in Runcorn, England, on a 1996 Volvo FL-12 chassis and is among a plethora of E-One apparatus delivered worldwide. The apparatus featured a massive 3000 gpm pump. It carried 200 gallons of water and 2000 gallons of foam in addition to a 1000-pound Hydro-Chem system. *Andrew Henry*

Although the mini-pumper craze had long ended, many departments still had a need for quick attack units. The Mohegan Tribal Fire Department in Montville, Connecticut, received this unit on a 1996 Ford F-550 4x4 chassis with a 250 gpm pump and 260-gallon water tank. *Mark Redman*

In the mid to late 1990s, E-One offered a Budget Buster pumper. It allowed a no frills pumper straight off the line with limited options and quick delivery. This stock model on a 1997 Ford F-750 chassis with 1250 gpm pump and 1000-gallon tank was purchased for the Mesa Community College Fire Program in Arizona. It was assigned S/O 18849. *David Greenberg*

Aboite Township in Fort Wayne, Indiana, received this pumper-tanker on a 1997 International 2674 chassis with four-door cab. S/O 18003 was equipped with a 1250 gpm pump and 3000-gallon water tank with jet dump on the driver side. *Kent Parrish*

In the mid to late 1990s E-One offered a full complement of economical commercial apparatus in its American Eagle program, not to be confused with the former company spun off by former E-One employees. Bryce Canyon National Park in Utah took delivery of this 1997 American Eagle 1800-gallon tanker with 300 gpm Waterous pump, which, although in the minority for E-One, was now an option. S/O 19761 also featured a diamond plated lower body. *Garry Kadzielawski*

Farmingdale, New York, operates this unique 4x4 rescue-pumper on an International 4900 chassis with 750 gpm pump, 250-gallon water tank, 250-gallon foam tank, and a light tower mounted on the cab roof. *Juergen Kiefer*

The Superior plant in Canada built a majority of the quick attack units and tankers under the E-One name from 1997 to its closing in 2006. Canadian-built E-Ones could be identified by the presence of both E-One and Superior (4-digit) S/O numbers. San Francisco, California, received this Super Lynx mini-pumper on a 1999 Ford F-550 4x4 chassis with a separate Volkswagen engine that powered a 650 gpm Waterous pump. It carried 300 gallons of water. *Garry Kadzielawski*

Throughout the later 1990s, a growing fire apparatus trend was larger brush apparatus to combat wildland fires and E-One was no exception. Eugene, Oregon, took delivery of this American Eagle wildland rig on a 1999 International 4900 4x4 chassis with 500 gpm pump, 500-gallon water tank, and 20-gallon foam cell. Note the steep angle of the rear body for encroachment. *Shane MacKichan*

The historic O.K. Coral and Tombstone, Arizona, are protected by S/O 121196, which was built on a 2000 Freightliner FL-80 chassis with 1250 gpm pump, compressed-air foam system (CAFS), and 1000-gallon water tank. From 2000 on, E-One added a "1" in front of its S/O sequence to differentiate apparatus from its sister company Saulsbury. Although this rig was identified as 121385, it was sequentially S/O 26476. Saulsbury apparatus fell under a six digit sequence as well, but with a "2" as the preceding number. *Don Feipel*

Parkers Mill, Kentucky, operates this 1250 gpm pumper with 1250-gallon water tank on a 2001 International 4900 chassis. This view shows S/O 123766 with its hydraulic ladder rack, a popular option that provided for more compartment space and larger water tanks. *David Mattingly*

This 1250 gpm pumper with 1000-gallon water tank was specified by Wynnewood, Oklahoma, on a 2002 Peterbilt chassis. Although fairly standard and not exceptionally large, S/O 124422 seemed much larger on the imposing choice of chassis. The Tradition program of economy apparatus had now replaced the American Eagle line. *Steve Loftin*

The E-One Jack Rabbit was a popular option for many departments in that it provided for a well-equipped brush unit or a combination quick response/rescue apparatus. Fort Knox, Kentucky, received two of these units from the Superior plant. S/O numbers 125469–70 were built on 2002 Ford F-550 4x4 chassis and featured 250 gpm PTO pumps, 250-gallon water tanks, and 25-gallon foam cells. *Kent Parrish*

The glitzy ski town of Park City, Utah, took delivery of this Wildcat urban interface rig in 2003 on the updated International 7400 chassis. The 4x4 unit featured a 500 gpm pump, 500-gallon water tank, 25-gallon foam cell, and a bumper mounted turret. *Stefan Farage*

Two of these massive pumper-tankers went to Klein, Texas. Built on 2003 Kenworth T-800 chassis, each featured a 2000 gpm pump and 3000-gallon water tank with a hydraulic folding dump tank rack on the driver side. The light tree above the first compartment contains large tank-level indicators. *Eric Hansen*

Although not prevalent, E-One did offer elliptical tankers in addition to the standard wet-side units. Compact Grizzly models offered under the American Eagle line and the Tradition series were built at the Superior plant. Gila River, Arizona, took delivery of this big Tradition tanker on a 2004 International 7600 chassis. S/O 127924 was equipped with a 500 gpm pump and 3000-gallon water tank. *Garry Kadzielawski*

Although front mount pumps were no longer common, St. Meinrad, Indiana, specified this unique rig on a 2004 International 7400 4x4 chassis with front mount 750 gpm pump and 1250-gallon water tank. Built at the Superior plant, it was assigned S/O 128820. *Kent Parrish*

This rig didn't have to travel far from the plant in Ocala to Flagler Beach, Florida. S/O 130257 was built in 2005 on a Freightliner M-2 chassis with four-door cab. The Tradition model featured a 1250 gpm pump and 1000-gallon water tank. *Kent Parrish*

In today's fiscally restrained fire service, delivery of a new Tradition 1250 gpm pumper with 1000-gallon water on a 2005 Freightliner M-2 chassis such as S/O 130756 for Mt. Eden, Kentucky, brings as much joy to towns across America as a Ford C model once did. From the innovative aluminum apparatus at its inception to the program apparatus of today, E-One has blazed a trail meeting the needs and wishes of fire departments big or small by providing a wide array of pumping apparatus and tankers on commercial chassis. *Kent Parrish*

Chapter 2

Emergency One began to offer apparatus on custom chassis around 1976. The body style for custom pumpers fell under the Protector IV designation. Seminole, Florida, received S/O 296 in 1976 on a Pemfab chassis, which was prevalent among early Emergency One custom apparatus. It was equipped with a 1250 gpm pump and 750-gallon water tank. *Garry Kadzielawski*

CUSTOM PUMPERS & TANKERS

With overnight success in quick attack and commercial pumpers, Emergency One began to field requests for apparatus on custom chassis. Without the capabilities to engineer and build its own early on, Emergency One utilized custom fire chassis from the likes of Duplex, Hendrickson, Pemfab, and Spartan to meet this demand.

Aided by the resources of Federal Signal, Emergency One unveiled its first municipal custom fire apparatus chassis in 1984. The original flat windshield and square cab combination of the Hurricane is considered a classic among modern fire apparatus buffs to this day. In 1985, a spin-off company named Federal Motors began operations in a building at the Ocala plant and was tasked with building the custom fire apparatus chassis for Emergency One. Although a separate subsidiary on paper, a Federal custom fire apparatus chassis was in essence an Emergency One custom chassis and will be assumed as such throughout this book. This arrangement paved the way for Emergency One to become an innovator among the custom fire chassis builders, from the revolutionary rear engine Hush through mid-range offerings such as the Protector to the flagship Cyclone II of today.

Along with the option of a custom chassis, larger capacity fire pumps became available. Morton Grove, Illinois, was the recipient of this Protector IV rescue-pumper on a low profile Hendrickson chassis with 1500 gpm pump and 500-gallon water tank. It was assigned S/O 666. Custom pumpers were still delivered within the industry leading 30–90 day time frame. *Chuck Madderom*

In addition to Class A pumpers, tankers became available on custom chassis with multiple pump and tank sizes. This big job was delivered to Oakbrook Terrace, Illinois, in 1978 on a Hendrickson chassis. S/O 756 featured a 2500-gallon water tank and 250 gpm PTO pump. *Chuck Madderom*

Emergency One delivered numerous custom pumpers on this style Spartan chassis with Maxi-Vision cab. South Holland, Illinois, received S/O 933 in 1979 with a 1500 gpm pump and 750-gallon water tank. During this year, Emergency One was acquired by the conglomerate Federal Signal, which would continue to bankroll operations and innovation as development began on an exclusive custom chassis. *Garry Kadzielawski*

Illinois was an early haven for custom Emergency One apparatus. Brookfield received this 1250 gpm Protector IV pumper with 500-gallon water tank in 1980 on a Hendrickson chassis. It was assigned S/O 1452. *Chuck Madderom*

Among the many apparatus loaned by Emergency One to production studios for use in film and on television was this 1500 gpm Protector IV pumper with 500-gallon water tank on a 1981 Pemfab chassis. S/O 1795 was a regular in the television show *Code Red*, before being sold to Roseville, California, and repainted from red to yellow. *Chuck Madderom*

The Hendrickson 1871 chassis with square cab was also a popular choice for early Emergency One custom apparatus. Annapolis, Maryland, was the recipient of this 1982 Protector IV equipped with a 1250 gpm pump and 500-gallon water tank. *Howard Meile*

The York Center Fire Protection District in Illinois took delivery of this big tandem axle job in 1983 on a Spartan chassis. S/O 2704 carried 3000 gallons of water in addition to a 1500 gpm pump. *Garry Kadzielawski*

In 1984, Emergency One introduced the Hurricane, the first municipal custom fire apparatus chassis of its own make. Built of ¼-inch aluminum, the original design featured a sloped front with square cab. Vertical side windows complimented a flat one-piece windshield. Riverside, California, received one of the first units in this 1500 gpm top-mount pumper with 500-gallon water tank. It was assigned S/O 3618. The Hurricane pumpers would also adopt the Protector IV body style designation with standard transverse compartment or optional top-mount control panel. *Chuck Madderom*

The Sentry IV body style was also designated in conjunction with the release of the Hurricane custom pumper. It was the custom extension of the Sentry III commercial body style with a standard pump panel with no transverse compartment. North Beach, Maryland, received this well-appointed Hurricane Sentry IV in 1986. S/O 4584 came complete with 1250 gpm pump, 1000-gallon water tank, mirror-polished pump panel, diamond-plate compartment doors, and wind deflector at the rear step. *Howard Meile*

The Kennedy Space Center in Florida received three of these Hurricane Protector IV 1500 gpm top-mount pumpers in 1984 with 35-gallon foam tanks and 100-pound dry chemical systems. Note the unique location of the booster reel. The units carried S/O numbers 3653-3655. Everything on the Hurricane was designed to be oversized, from the windows to the jump seats to the engine air intakes. *Garry Kadzielawki*

The city of Chicago found early favor with Emergency One in aluminum pumpers and aerials on commercial and proprietary chassis. Beginning in 1985, the department began to order custom Hurricane Sentry IV pumpers in bulk. They typically featured 1250 gpm pumps, 500-gallon water tanks, and the telltale Chicago bumper extension with accordion hose load. This order contained S/O numbers 4265-4281. *Garry Kadzielawski*

Another obvious trait of the early Hurricane custom chassis was the presence of large door and canopy cab windows that extended nearly to the wheel well. Tulsa, Oklahoma, operated two identical 1985 Protector IV models with 1250 gpm pumps and 500-gallon water tanks with S/O numbers 4398 and 4399. During this year, Federal Signal spun off Federal Motors on the grounds of the Ocala plant to build custom fire apparatus chassis for Emergency One and commercial applications. Often the builder plate inside the cab would state the apparatus as an Emergency One on a Federal Motors chassis. But, in essence, it was an Emergency One custom. *Steve Loftin*

In 1985, Emergency One introduced its second custom offering, the revolutionary Hush rear-engine chassis. By placing the engine at the rear of the apparatus, the enclosed, four-door cab provided a quiet and spacious open interior for seating up to 10. The first production model, this demonstrator was purchased by Wyandanch, New York. It featured a 1500 gpm pump and 500-gallon water tank. *John Toomey*

This strikingly painted Hurricane Protector IV pumper was fittingly delivered to Bayou Blue, Louisiana, in 1986. S/O 5091 featured a 1250 gpm pump with top-mount controls and 1000-gallon water tank. *Glenn Vincent*

Soon after the introduction of the original square cab Hurricane in 1984, a contour cab version was made available. This contour Hurricane Protector IV with four-door, fully enclosed cab was specified by Chambersburg, Pennsylvania, in 1986. The industry standard for enclosed cab configurations would be SMFD (short cab), MFD (medium cab), LFD (long cab), and XLFD (extended cab). It was equipped with a 1250 gpm pump and 750-gallon water tank. *Joel Woods*

The Emergency One U.S. One budget pumper was also available on the Hurricane chassis with the telltale lack of characteristic diamond plate around the rear wheel well. Crumpton, Maryland, received this 1986 model with Protector IV body style. It was equipped with a 1500 gpm pump and 750-gallon water tank. *Joel Woods*

The Hush was available with a wide array of cab and body arrangements. East Berlin, Connecticut, took delivery of S/O 5227 in 1986 with an SMFD (short) cab. It was equipped with a 1500 gpm pump and 750-gallon water tank. The Hush enclosed cab was the first Emergency One model to feature advancements such as compartment-open warning lights and creature comforts such as air-conditioning. *Mark Redman*

This view displays the design concept of a Hush rear-engine chassis. The engine compartment was placed to the rear of the water tank and underneath the hose bed. This unit was one of five pumpers with SMFD cabs purchased by Anaheim, California, in 1986 with 1500 gpm pumps and 500-gallon water tanks. They were assigned S/O numbers 4918-4922. *Chuck Madderom*

Emergency One continued to produce apparatus on proprietary custom chassis, including the venerable Mack CF chassis, as Mack was no longer producing complete fire apparatus. Lyndon, Kentucky, received this Protector IV on a Mack CF chassis in 1986. S/O 4706 was equipped with a 1500 gpm pump and 500-gallon water tank. *David Mattingly*

Under the guidance of Federal Motors, Emergency One continued its thrust in the custom market with the 1986 introduction of the Cyclone custom chassis. Similar in principle to the Hurricane, it featured a taller and more prominent cab with a noticeable gap between the side windows and roof drip line. The contour cab had smaller side windows compared to the Hurricane. Pleasantview, Illinois, received one of the first Cyclones. S/O 7747 featured a 1500 gpm pump and 750-gallon water tank. *Garry Kadzielawki*

This Cyclone was delivered to Canton, Connecticut, in 1987 with a 1500 gpm pump and 500-gallon water tank. It was assigned S/O 5314. The Sentry IV and Protector IV custom body styles had also been discontinued in favor of either a Protector XL or Protector XLT designation for commercial and custom pumpers. The Protector XL had a side-mount pump panel. *Glenn Vincent*

The new Protector XLT pumper designation indicated the apparatus was equipped with a top-mount control panel, as in this 1989 Cyclone for Terryville, Connecticut. S/O 7946 featured a 1250 gpm pump and 750-gallon water tank. By now, Emergency One had become known simply as "E-One" and had updated its builder plates on the apparatus in reflection. *Mark Redman*

The heavy-duty Hurricane chassis also made it an option for departments in the market for a big custom tanker such as this 1987 model for Hebron, Connecticut. S/O 5234 was equipped with a 1000 gpm pump, top-mount controls, and 2000-gallon water tank. *Mark Redman*

Despite its high cost, the advantages of the Hush rear-engine chassis appealed to departments, big and small. The small town of Springfield, Tennessee, received S/O 7225 in 1987 with an SMFD cab, 1500 gpm pump and 500-gallon water tank. The Hush body now received the Hush XL designation with various configurations available in relation to the rear-engine design. *Kent Parrish*

Many of the larger crash/fire/rescue departments run structural pumpers to better protect their airports. Miami International Airport operated this 1988 Cyclone Protector XL pumper with LFD cab, 1500 gpm pump, 500-gallon water tank, and 20-gallon foam cell. It was assigned S/O 5683. An option on the Cyclone was full-length side windows, which gave a Hurricane appearance. *Bob Milnes*

E-One combined with Fire Trucks Inc. (FTI) to provide many of these structural pumpers for the military. They featured Cyclone chassis and cabs with full-length driver and passenger Hurricane windows. FTI built the bodies with 1000 gpm pumps, 500- gallon water tanks, and 40-gallon foam cells. This 1989 model went to Fort Meade, Maryland. *Joel Woods*

Despite a near dominance in the single-source custom market, E-One would still build on other custom chassis per customer request. Columbia, Pennsylvania, preferred a Duplex Vanguard chassis for this 1989 pumper equipped with a 1500 gpm pump and 750-gallon water tank. It was assigned S/O 7064. *Rick Rudisil*

Claymont, Delaware, received this well-equipped Hush XL rescue-pumper in 1989 with a 10-man LFD cab and extended body. S/O 6740 featured a 1500 gpm pump and 750- gallon water tank. *Frank Wegloski*

All of E-One's custom chassis were available with four-wheel-drive, an option Bristol, Connecticut, chose on this 1990 Cyclone, which was identifiable by the higher stature of the LFD cab. S/O 8439 was equipped with a 1500 gpm pump, 750-gallon water tank, and 20-gallon foam cell. *Mark Redman*

North Attleboro, Massachusetts, received this Cyclone in 1990 with LFD cab. S/O 6727 featured a roof-mounted turret and carried 100 gallons of foam in addition to 750 gallons of water and a 1250 gpm pump. In 1990, E-One discontinued the XL and XLT pumper identifications and began to simply designate apparatus as commercial or custom pumpers. *Joel Woods*

Phoenix, Arizona, was nicknamed the "Hush Capital." The department purchased approximately 63 E-One Hush pumpers from 1987 to 2000, including an experimental rig with independent front suspension. The specifications varied from order to order, but included side- and top-mount pump panels and some with booster reels mounted in the front bumper. Eleven were delivered in 1990 with 1500 gpm pumps, 500-gallon water tanks, and 40-gallon foam cells that featured air-actuated, bifold "bus-style" crew doors. *David Greenberg*

The E-One Hush was available with an option that allowed the rear compartments to slide back, providing access to the rear-mounted engine. This capability could be differentiated by the top line of the rear upper compartment being slightly off from the center and forward compartments. The Hush Vista included a raised roof cab with enclosed, top-mount pump panel. Sun City West, Arizona, received S/O 8252 in 1990 with a 1500 gpm pump and 500-gallon water tank. The Vista roof was unique in that it sat directly on top of the cab and slightly in, so as to not comprise the aluminum extrusions of the cab shell. The Vista roof would also be available on future E-One tilt-cab models. *David Greenberg*

In 1990, E-One added its fourth custom chassis to the lineup. Named after the discontinued body style designation, the Protector was an entry-level model that featured an SMFD four-door fixed cab of engine forward design with a low engine tunnel or "doghouse." Standard was the smaller Cummins 6CTA 8.3 engine with 250-300 horsepower. Goldsboro, Maryland, received one of the first deliveries. It was equipped with a 1250 gpm pump and 1000-gallon water tank. *Howard Meile*

The Protector provided an affordable fully custom apparatus. The Storrs campus of the University of Connecticut was protected by this 1990 model equipped with a 1250 gpm pump and 750-gallon water tank. It was assigned S/O 7615. E-One often, but not always, began to differentiate custom chassis with an emblem above or below the middle cab window. With additional options and the highest horsepower engines available on this chassis, some Protectors actually escaped Ocala with Cyclone nameplates, which added confusion to the model lineup. *Mark Redman*

Kankakee, Illinois, received this Cyclone in 1990 with LFD cab. S/O 8915 was equipped with a 1500 gpm pump, 500-gallon water tank, and 30-gallon foam cell. The apparatus featured diamond plating on the lower compartment doors. NFPA would soon outlaw firefighters riding on the outside of the apparatus and require seat-belted positions for each firefighter. All new fire apparatus would be required to have a fully enclosed cab. *Garry Kadzielawski*

In 1991, the Protector was redesigned as a true engine forward model and became E-One's first tilt-cab offering. It also became available in longer cab lengths. The tilt-cab was identifiable by the presence of a front grill and small side air vents that followed the contour of the front wheel wells. Venice, Florida, received this Protector TC (tilt-cab) LFD in 1991 with a 1250 gpm pump and 750-gallon water tank. It was assigned S/O 8926. *James Brown*

With the Cyclone and new Protector dominating the pumper line at Ocala, new Hurricane pumpers were becoming fewer. However, New Paltz, New York, preferred the Hurricane for this 1991 rescue-pumper, which featured a fully enclosed body with hidden pump panel. S/O 8925 was equipped with a 1250 gpm pump and 500-gallon water tank. *Neal Van Deusen*

This Hush pumper with SMFD cab was delivered to Washington Township at Centerville, Ohio. S/O 8955 featured a 1500 gpm pump and 500-gallon water tank. The Hush emblem was displayed below the middle cab window. *Kent Parrish*

Later in 1991, the Cyclone was also made available as a tilt-cab, which would become the new flagship of the E-One custom line. The largest engines, pumps, and water tanks were available. Although the Protector was now only available as a tilt-cab, the original fixed-cab Cyclone would still be available and could be identified by the lack of a front grille and vertical side air vents. Andover, Connecticut, received this Cyclone TC (tilt-cab) in 1991. S/O 8977 had a 1500 gpm pump and 1000-gallon water tank. *Shaun P. Ryan*

In 1992, Federal Signal purchased Superior Emergency Vehicles of Canada. Up until 1988, E-One had sold approximately just a dozen units in Canada. HUB Fire Engines became the E-One Canadian dealer in 1988 and sold around 40 units until the purchase of Superior. Superior would continue to satisfy its Canadian customers in addition to building bodies on E-One custom chassis. View Royal in British Columbia received this Superior pumper on a 1992 E-One Protector TC chassis with a 1250 gpm pump and 600-gallon water tank. *Dave Stewardson*

Fishers, Indiana, received this Cyclone TC with MFD (medium) cab in 1992 with a unique paint scheme and fire prevention graphics. S/O 10952 was equipped with a 1500 gpm pump and 1000-gallon water tank. The Cyclone TC and Protector TC were often difficult to differentiate at quick glance. Generally the Cyclone TC would have a larger front grill and windshield wiper arms that originated just above the headlight bezels. The Protector TC typically had a smaller front grill and windshield wipers installed just below the windshield. However, a customer could order options for a Protector TC that could make it nearly identical to a Cyclone TC aesthetically. The Protectors did have different interior layouts that set them apart upon further examination. *Frank Wegloski*

Long Green, Maryland, chose the Cyclone TC LFD chassis with its heavier frame and larger engine capability for this big green job equipped with a 1500 gpm pump and 2000-gallon water tank. Of note during this period was the use of Godiva pumps in select E-One apparatus as the European brand had been purchased by Hale and eventually fell under control of Hale Products Europe. During this period, E-One apparatus were equipped with a mixture of Hale and Godiva brand pumps. *Warren Jenkins*

E-One modernized its old ambulance-pumper concept on the Hush chassis. Sandy, Utah, received this EMS/transport-capable, 1500 gpm pumper with 500-gallon water tank in 1993. The open XLFD (extra-long) cab interior with Vista roof allowed room for treatment while transporting a patient on a stretcher. *Stefan Farage*

Lighter duty than the Cyclone TC, the Protector TC was generally available only in a single rear axle configuration. Eaton, Ohio, took delivery of this Protector TC with SMFD cab in 1993. The 1250 gpm unit carried the maximum of 1500-gallons of water on the Protector TC chassis, which was optimized by deleting upper side compartment space. It carried S/O 12409. *Shaun P. Ryan*

Dating back to the mid 1970s when Emergency One was awarded a contract to built riot vehicles for Nigeria, it had established an international presence. The Wonderboom suburb of Pretoria, South Africa, received this Cyclone TC MFD in 1994 with Vista roof and enclosed top-mount pump panel. In this configuration, the Vista roof extended out from the basic cab shell and partially over the pump. S/O 13249 was equipped with a 4725 gpm pump (Imperial), 3700-liter water tank, and 450-liter foam cell. *Shaun P. Ryan*

The resort area of Lake Placid, New York, took delivery of this 4x4 Hurricane pumper in 1994. S/O 13923 featured a 1500 gpm pump and 1000-gallon water tank with dual 30-gallon foam cells. The Hurricane had also been made available as a tilt-cab, which was more of a request on the aerial apparatus. A Hurricane TC could be identified by the presence of two slots between the cab and front bumper, which allowed for the cab to tilt up, but otherwise, was virtually indistinguishable from a fixed cab Hurricane at a distance. *Neal Van Deusen*

In 1993, E-One introduced its fifth custom chassis. The Sentry was another entry-level offering with a Cummins 6CTA engine, but situated below the Protector TC in hierarchy. The tilt, MFD cab featured as standard front grill and side intakes cut directly into the cab skin and painted to match. The cab doors were also flush with the cab and angular, as opposed to the standard lap style utilized by E-One that followed the curvature of the wheel well on the other models. The private Rural Metro Fire Department purchased S/O 13212 in 1994 with 1250 gpm unit, 1000-gallon water tank, and 30-gallon foam cell to protect Paradise Valley, Arizona. *David Greenberg*

Riverside County, California, received fifteen Sentry pumpers in 1994 with S/O numbers 14101 through 14116. They were upgraded to feature chrome front grills and side intakes. The units were equipped with 1500 gpm main pumps and 250 gpm auxiliary pumps for pump and roll operations. They carried 500 gallons of water and 30 gallons of foam. The introduction of the Sentry also coincided with the availability of Waterous pumps on E-One apparatus for the first time. *Chuck Madderom*

E-One introduced one of the largest industrial pumpers to date in the Liquidator based on a Hush chassis. The Marathon-Ashland Oil refinery in Catlettsburg, Kentucky, purchased this 1995 demonstrator model with S/O 14223. The Liquidator featured an MFD cab with modified Vista roof. A Williams Hot Shot II foam system with 750-gallon tank worked in conjunction with a 6000 gpm Allis Chalmers pump and Hydro-Foam deck gun. *Greg Stapleton*

Fairfax County, Virginia, had long been a loyal E-One customer. Engine 12 was assigned this Protector TC pumper in 1995 with MFD cab. It was equipped with a 1250 gpm pump, 750-gallon water tank, and 40-gallon foam cell. *Joel Woods*

In 1996, New York City evaluated an experimental E-One pumper that was built on a Hush chassis with modified MFD cab and featured a Mack engine. Not to be confused with several loaners marked FDNY for television shows over the years, this was the first and only E-One pumper operated by the FDNY to date. The 1000 gpm unit with 500-gallon tank served brief tours at four different engine companies before being turned in. It was assigned S/O 15176. *John Toomey*

Seattle, Washington, was another loyal E-One customer from the late 1990s on. This 1996 Protector TC pumper with MFD cab and Vista roof featured a 1500 gpm pump with top mount controls and 500-gallon water tank. It was assigned S/O 16209. *Bob Dubbert*

This Sentry pumper was delivered to Laurel County, Kentucky, in 1996. S/O 16173 was equipped with a 1500 gpm pump and 1000-gallon water tank. Most departments did not prefer the standard features of the Sentry and opted for the chrome front grill and side air intakes. *David Mattingly*

In 1996, E-One discontinued the original Cyclone series, the Protector TC, and the short-lived Sentry. Their production runs ended once the backlog was completed. They were replaced with a new flagship, the redesigned Cyclone II. The new model could be differentiated from the old Cyclone and Protector models by a larger front grill and the elimination of side air intakes. The standard MFD cab interior was updated to be more ergonomic and open. Springfield, New Jersey, received the first delivery, this 1500 gpm unit with 700-gallon water tank and 50-gallon foam cell. *Scott Mattson*

The replacement of three models with the Cyclone II effectively eliminated E-One from the entry- to mid-level custom market, so it introduced a custom offering in its economic American Eagle program later in 1996. In essence, the American Eagle custom cab was a continuation of the basic Protector TC chassis and LFD cab shell. E-One also updated its raised roof design to be completely integral with the aluminum extrusions of the cab shell, rather than separate as the old Vista style was. North Little Rock, Arkansas, received this American Eagle custom in 1996. S/O 16512 had a 1250 gpm pump and 750-gallon water tank. *John Arthur*

The Hurricane was slowly fading into history in the E-One custom pumper lineup and later models were becoming rare. Spotsylvania, Virginia, did choose a Hurricane TC chassis for this nice looking 1250 gpm pumper with 750-gallon water tank in 1997. It featured a raised roof cab, which was uncommon for the Hurricane. The Hurricane had also adopted the standard MFD cab configuration. *Joel Woods*

Although the majority of American Eagle custom pumpers were rather plain and ordinary, the cost savings associated with lighter chassis and smaller engines allowed some departments to accessorize quite well. Desoto, Missouri, received this nice looking unit in 1997 with a 1500 gpm pump and 750-gallon water tank. It was assigned S/O 18239. *Dennis J. Maag*

The Motiva Norco refinery in Louisiana is protected by this massive industrial pumper based on a 1998 Cyclone II chassis. S/O 18475 features a 3000 gpm pump with top mount controls, rear-mounted deluge gun, and light tower on the cab roof. The apparatus carries 500 gallons of water and 2000 gallons of foam. *Joel Woods*

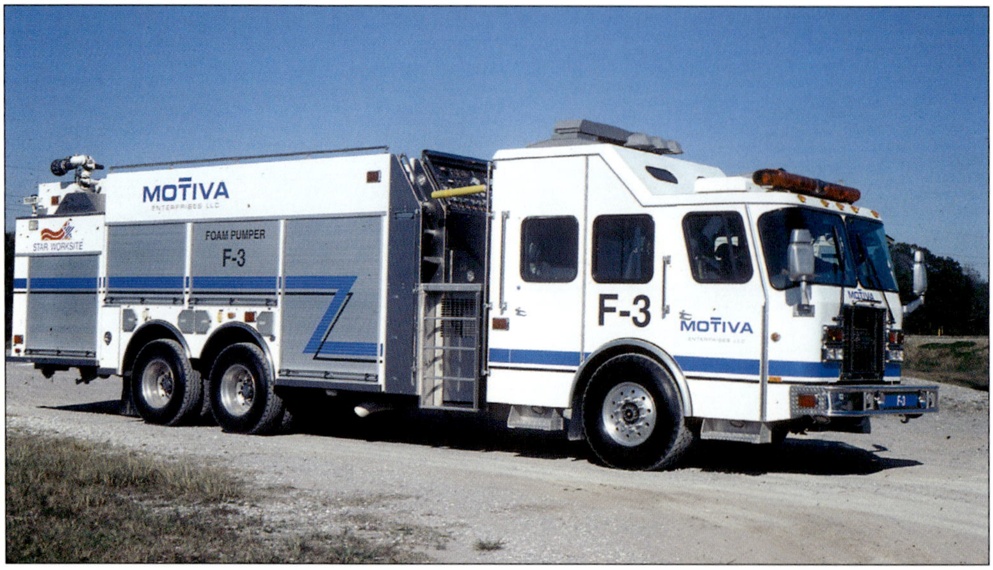

In 1997, E-One introduced a space-age Concept 2000 rescue-pumper. Developed through an alliance with the plastics industry, the aerodynamic vehicle was made of an aluminum shell covered by a skin of a composite polymer called Millennium III. It was renamed the Daytona and entered production in 1999. A unique feature was front doors that hinged on the "B" post. A 1500 gpm rear-mounted pump, 500-gallon water tank, and 30-gallon foam system were hidden within the fully enclosed body. Approximately ten units, most of which have been scrapped or sold by their original customers, were produced before the Daytona was discontinued. S/O 19944 is currently serving Garrison, Kentucky, via East Lake, Florida. *Kent Parrish*

By 1999, the entry-level American Eagle custom pumper had been replaced by what E-One called a "super-commercial" on a proprietary chassis. The Sentry II was a standard 1250 gpm unit with a 500-gallon water tank placed on a Spartan Advantage chassis and cab with optional E-One trim. S/O 120680 served as a loaner and demonstrator model. *Chuck Madderom*

Cahokia, Illinois, received this impressive rescue-pumper in 2000 on a Cyclone II chassis. The bin on top of the raised roof cab contained a light tower. It was equipped with a 1500 gpm pump, 1000-gallon water tank, dual 50-gallon foam cells, and four-bottle air cascade system. With "1" now appearing in front of the S/O sequence to differentiate an E-One from a Saulsbury, the apparatus was identified as 121385. *Dennis J. Maag*

In an effort to boost its ISO insurance ratings, Spring, Texas, received thirteen new E-One apparatus in 2001, including two of these monsters with 3000 gpm pumps and 3500-gallon water tanks on Cyclone II chassis. The new purpose-minded fleet was designed to meet any need by bringing every available option to any given scene. *Eric Hansen*

E-One again changed its entry-level custom offering, this time to coincide with its new Tradition economy program. Still available was a Spartan Advantage chassis and cab. Sellersburg, Indiana, received two of these units in 2001 with 1250 gpm pumps and 1000-gallon water tanks in the department's white over blue scheme. They were assigned S/O 123095 and 123096. *Kent Parrish*

In 2001, with an empty space in the lineup for a mid-range custom, E-One introduced a breath of fresh air and a return to weather-themed models with the Typhoon custom chassis. With a cab shell similar to the Cyclone II, it offered a smaller drive train and engine choices to keep the cost down. Distinguishable features included an updated front grill with painted shell and the model name centrally integrated into a horizontal section along with a return to side air vents that followed the contour of the front wheel well. East St. Louis, Illinois, received one of the first deliveries. S/O 124460 had a 1500 gpm pump and 1000-gallon water tank. *Dennis J. Maag*

The new Typhoon became an immediate hit as it provided a choice for departments that fiscally did not need or could not afford a loaded Cyclone II. By downgrading to a Typhoon, which was included in the Tradition economy program, many departments were able to order multiple units. Kingston Springs, Tennessee, received two Typhoon Tradition pumpers in 2002 equipped with 1250 gpm pumps and 1000-gallon water tanks. *Kent Parrish*

Saulsbury was not only a pioneer in stainless fire apparatus bodies, but was an innovator in unique designs. With a growing trend in such apparatus, E-One teamed up with Saulsbury to build a stainless steel rear-mount rescue-pumper. Spanish Lake, Missouri, received this Saulsbury on a Cyclone II chassis in 2003 with a 1500 gpm pump, 770-gallon water tank, and 50-gallon foam cell. It carried E-One S/O 126874. *Dennis J. Maag*

By 2003, E-One had redesigned the fading Hush models as a mid-engine format with rear-mount pump. The design included a modified Cyclone II cab and Saulsbury stainless steel body. It still provided a quiet and open cab interior, but with a more ergonomic rear end and better weight distribution. This 2003 Hush ME (mid-engine) went to Frisco, Texas. S/O 127575 had a 1500 gpm pump, 500-gallon water tank, and dual 30-gallon foam cells. *Eric Hansen*

E-One has continued to have a presence in the international and industrial market. This 2003 Hurricane industrial pumper was delivered to the BP Saltend chemical plant in Hull, England. The 3000 gpm unit carried 500 gallons of water and 500 gallons of foam. The custom chassis that pushed E-One to the forefront would see limited production over the next few years as market demand for this format all but disappeared and new EPA emissions standards eventually spelled the demise of the venerable Hurricane by 2007. *Andrew Henry*

The new Hush ME proved to be popular with longtime Hush customers, particularly out west and including fourteen units for Phoenix. The stainless steel body was believed to have many of the same benefits of aluminum, but last even longer. Glendale, Arizona, received S/O 127301 in 2004 with a 1500 gpm pump, 500-gallon water tank, and combined 50-gallon foam cells. Also suffering from a lack of market acceptance, the final death knell for the Hush also came with the 2007 EPA emissions standards. *David Greenberg*

Although the Cyclone II is considered E-One's flagship custom chassis, the value of the Typhoon has practically pushed it into the sales lead. Lynnville, Indiana, received this nice looking top-mount rescue-pumper in 2005. S/O 129687 was equipped with a 1250 gpm pump and 1000-gallon water tank. *Kent Parrish*

Not a color commonly seen on fire apparatus, Norfolk, Massachusetts, received this mean looking black Typhoon in 2005. S/O 129259 featured a 1500 gpm pump and 750-gallon water tank. The standard painted Typhoon front grill could be upgraded with chrome, adding to its appearance. *Dick Bartlett*

By 2003, Superior had become E-One Canada with its deliveries having bodywork identical to that of E-One and the Superior name within a similar oval builder's plate. Prince Albert in Saskatchewan, Canada, received this Superior on a 2006 E-One Cyclone II chassis with S/O 128542. It featured a fully enclosed top-mount panel for the 1250 gpm pump and carried 500-gallons of water with dual 50-gallon foam cells. From 1992 to 2006, E-One directly sold approximately 50 rigs to Canadian departments while the vast remainder was completed by Superior. *Dave Stewardson*

Boston had long been an E-One customer, from commercial pumpers to aerials to specialty units. After a trial period with another manufacturer, Boston returned to E-One in 2007 for a batch of Cyclone II 1250 gpm pumpers with 500-gallon water tanks. The Cyclone II had also undergone its first cosmetic upgrades, complete with a new front grill, similar in appearance to that on the Typhoon, but a bit fancier. *Joel Woods*

As the first decade of the new millennium began, E-One began to show signs of suffering from management and quality control issues through a stagnant product line and sluggish sales. Despite the gloom, a bright spot in 2007, literally, was this Cyclone II rear-mount rescue pumper built for the Pocono Summit Volunteer Fire Company in Pennsylvania. It was equipped with a 1500 gpm pump and 500-gallon water tank. *Patrick Rudisil*

In 2007, E-One again gave a look into the future with the introduction of the all new Quest custom chassis. Designed with firefighter safety and ergonomics in mind, it was hoped to reinvigorate a stagnant product line and display its renewed dedication to innovation. With a streamlined custom lineup of the Typhoon, Cyclone II, and Quest, E-One begins its "quest" to reclaim its position as an industry leader. *Dick Logan*

Chapter 3

Hale Fire Pumps wanted to provide an aerial ladder to builders that did not manufacture their own and selected Emergency One, its largest client, as an OEM builder for the Fire Spire 106-foot ladders from National Foam, a sister company. Pleasantville, New Jersey, received this "Emergency One06" on a 1979 Pemfab chassis. The Fire Spire was a four-section, steel aerial. *Scott Mattson*

AERIAL APPARATUS

The late 1970s marked Emergency One's first venture into the aerial market when they obtained use of the Fire Spire aerial ladders from National Foam, which was a sister company to Hale Fire Pumps. In 1977, Hahn Motors acquired the design rights to the Fire Spire. After failing to purchase these rights from Hahn in 1979, Emergency One hired the engineer who originally designed the Fire Spire and worked with companies in the aluminum industry to help in the development of its own aerial line. The first aerial device, the 55-foot Strato-Spear telescoping waterway, was introduced in 1980. Emergency One also had early access to articulating and telescoping aerial devices from other manufacturers such as Simon and Snorkel.

In 1981, Emergency One introduced its first exclusive aerial ladder. The all-aluminum 110-foot StratoSpear aerial was followed two years later by the 95-foot StratoSpear aerial platform. To date, the aerial line has grown to include a full range of aerial products, including the Bronto Sky Lift.

Richardson, Texas, operated this quint on a low-profile 1980 Hendrickson chassis with a 106-foot Fire Spire. S/O 1038 was equipped with a 150 gpm pump and 250-gallon water tank. The Fire Spire was also made available as an 85-foot model. *Eric Hansen*

In addition to the Fire Spire, Emergency One had access to the Snorkel product line. Huntingburg, Indiana, took delivery of this 50-Telesqurt on a 1980 Hendrickson chassis with 1250 gpm pump and 500-gallon water tank. *Frank Wegloski*

Emergency One was the first American distributor of the British-built Simon Snorkel which was marketed as the StratoSpear 103-foot elevating platform. The three-boom, rear-mounted, articulating aerial device had a basket at the end and an aluminum ladder along the boom sections with a transfer stand between the two main booms. S/O 1667 was built for St. Lucie County, Florida, on a low profile Hendrickson chassis with a 1500 gpm pump and 300-gallon water tank. Approximately 20 of these units were built, with several special 77-foot units going to the Middle East. *Garry Kadzielawski*

Lauderdale Lakes, Florida, received S/O 2000 in 1981. The 106-foot Fire Spire was built on a Pemfab chassis with a 1500 gpm pump and 200-gallon water tank. After failing to purchase the design rights to the Fire Spire from Hahn Motors, which had acquired the design rights from National Foam, Emergency One began research and development of its own exclusive line of aluminum aerials. Approximately fifteen Emergency One Fire Spires were built. *Glenn Vincent*

Emergency One introduced its first exclusive aerial device in 1980 with the 55-foot StratoSpear telescoping waterway after modifications were made to a 65-foot prototype. Similar in concept to the Snorkel Telesqurt, it featured box boom construction with dual waterways on the exterior and remote controlled master stream. A rescue ladder was attached to the top of the device. Livermore, California, received S/O 2255 on a 1981 Hendrickson chassis with a 1500 gpm pump and 300-gallon water tank. *Chuck Madderom*

After two years of development with the original designer of the Fire Spire and guidance from companies in the aluminum industry, Emergency One introduced its first exclusive aerial ladder in 1981. The four-section, all-welded extruded aluminum ladder was called the StratoSpear 110-foot aerial ladder. Although thicker than the Fire Spire, it was roughly 1/3 lighter in weight and gave an extra four feet of reach, which made it the tallest ladder available in America at the time. The first unit was S/O 1743 on a Hendrickson chassis purchased by St. Augustine, Florida. *Joel Woods*

The StratoSpear featured a 200-pound tip load and had an optional waterway. The original design was based on a torque box integral to the Hendrickson chassis frame, as opposed to the industry standard of a bolt-on design. Stabilization was provided by four underslung, scissor or crisscross style jacks, a system based on a design more common to the crane industry that provided a lower center of gravity and narrower jack spread. S/O 2498 was delivered to Orange County, Florida, in 1982 on a Hendrickson chassis with 1250 gpm pump and 300-gallon water tank. *Glenn Vincent*

In 1983, Emergency One followed up with the introduction of the StratoSpear 95-foot platform. It was a three-section ladder with an 800-pound capacity platform that measured fifteen square feet with the same stabilizing system, but a slightly increased jack spread. The first units were built on Pemfab Maxi 944-T chassis, including S/O 3151 for Gurnee, Illinois, with a 1500 gpm pump and 200-gallon water tank. *Garry Kadzielawski*

The unusual Pemfab Wedge chassis was utilized on many fire apparatus, including this StratoSpear 110-foot ladder for Albuquerque, New Mexico. S/O 3069 was delivered in 1983. *Eric Hansen*

Also in 1983, Emergency One introduced the StratoSpear 135-foot aerial ladder, which overtook the original 110-foot model as the tallest ladder available in America. It was a five-section ladder with a 200-pound tip load with the same jack spread as the 95-foot platform. A pre-piped waterway was available at the tip of the fourth fly section. S/O 3128 was built on a 1983 Pemfab Royale chassis and delivered to Chicago, which had become a repeat Emergency One aerial customer. *Chuck Madderom*

The Hendrickson chassis was also available for the aerial platforms. Santee, California, received this StratoSpear 95-foot platform in 1983. S/O 2912 was equipped with a 1500 gpm pump and 200-gallon water tank. Emergency One had created five aerial sections that were all compatible with each other, making it possible to introduce ladders and platforms of various lengths. *Chuck Madderom*

Boston also became an early repeat Emergency One aerial customer and to date has used them almost exclusively. In 1984, Emergency One introduced the Hurricane, its first single-source, custom municipal fire apparatus chassis. A group of StratoSpear 110-foot aerial ladders with S/O numbers 3739-3755 were delivered to Boston in 1984 on the new Hurricane chassis. *Joel Woods*

The StratoSpear 55-foot telescoping waterway was available on commercial and custom chassis. This one-of-a-kind unit was delivered to Bayou Cane, Louisiana, in 1984 on an International CO chassis with canopy cab. S/O 2875 was equipped with a 1250 gpm pump and 300-gallon water tank. *Glenn Vincent*

Emergency One's StratoSpear 80-foot aerial ladder provided a compact unit, especially for single rear axle quint apparatus. The hometown Ocala Fire Department drove right down the street to pick up S/O 3510. It was built on a 1984 Hurricane chassis with 1250-gpm pump and 300-gallon water tank. *Joel Woods*

The Pemfab custom chassis was also utilized for early Emergency One aerials ladders. Odenton, Maryland, received this Strato-Spear 110 in 1984. S/O 3360 featured a unique, forward facing, supplemental crew compartment on each side of the body. *Joel Woods*

The Franconia Volunteer Fire Department in Fairfax County, Virginia, took delivery of this StratoSpear 95-foot aerial platform in 1985 on a Hurricane chassis with fully enclosed cab and S/O number 3835. *Joel Woods*

In 1986, by the request of dealers that represented older, traditional areas, Emergency One introduced the StratoSpear 96-foot mid-mount aerial ladder. Maynard, Massachusetts, received S/O 4433, the pilot model on a Hurricane chassis. The product didn't prove to be successful and just one more was built. The StratoSpear designation for the aerials had been discontinued and the models simply fell under E-One Aerial Systems. *Joel Woods*

Westminster, California, purchased this versatile apparatus with a 55-foot telescoping waterway on a 1986 Hurricane chassis. S/O 4481 was equipped with a 1500 gpm pump and 500-gallon water tank. Although aerials now simply fell under the E-One name, these units were often still called StratoSpears and even marketed as Telebooms or Telescopic Booms. *Chuck Madderom*

California continued to be a strong sales territory for Emergency One. This StratoSpear 95-foot aerial platform was delivered to Buena Park, one of many repeat west coast customers, in 1986. S/O 4492 was equipped with a 1500 gpm pump and 200-gallon water tank. *Chuck Madderom*

The Nation's capital was also a heavy user of Emergency One apparatus. S/O numbers 4928 and 4929 were delivered in 1987 on Hurricane chassis with forward facing, jump seat enclosures recessed into the body and 110-foot aerial ladders. The StratoSpear designation for the aerials, with the exception of the 55-foot telescoping waterway, had been discontinued and the models were designated as simply E-One. *Joel Woods*

Chicago was famous for being the first American fire department to use an articulating platform, or "Snorkel," back in 1958. Emergency One was part of a new generation of the concept for the city. Snorkel Squad 1 was assigned this compact 55-foot Snorkel on a 1987 Spartan chassis with fully enclosed cab. It carried S/O 5004. *Garry Kadzielawski*

The 135-foot aerial ladder had proved to be popular with many large cities and even smaller villages, such as Westmont, Illinois, which received S/O 4898 in 1987 on a Hurricane chassis. The reach of the 135 was unparalleled in the American market. By now, Emergency One was simply known as just "E-One." *Garry Kadzielawski*

Murrieta, California, received a fleet of Hurricane apparatus from E-One in this two-tone paint scheme, including this 95-foot platform in 1988. While most customers now preferred the contour version of the cab, this department preferred the square cab. The platforms had been redesigned with waterway piping becoming integral support and basket space increased by one square foot. S/O 6599 featured a big 1750 gpm pump and 200-gallon water tank. *Chuck Madderom*

Fairfax County, Virginia, was also a heavy user of E-One apparatus, including a fleet of aerials such as this 1989 delivery with a 110-foot aerial ladder on a tandem axle Hurricane chassis assigned S/O 6316. Fairfax specified the driver and officer windows to be shorter, giving an offset look with the full-length Hurricane windows on the crew cab windows. It can be noted that the fully enclosed aerial cabs could also be designated according to cab length (short, medium, long, extra-long) as the pumpers were. *Joel Woods*

By 1988, E-One had developed a telescoping waterway/aerial that could be placed on the rear-engine Hush chassis. The 50-foot Tele-boom was available in a mid-mount format, which made it applicable for the Hush design. This configuration placed access to the device between the cab and pump panel. Running Springs, California, received this unique 4x4 Hush in 1989. S/O 7398 featured a 1500 gpm pump and 500-gallon water tank. *Juergen Kiefer*

The Reedy Creek Improvement District, which protected the fantasy land of Walt Disney World in Florida, specified this unique apparatus on a tandem axle, all-wheel-drive Hurricane chassis in 1990. S/O 7438 was equipped with a 250 gpm PTO pump and 200-gallon water tank, but its primary intent was off-road rescue from the elevated monorail system that circled Disney World. The modified aerial ladder had a reach of 62 feet. *Stefan Farage*

E-One had built up a miniscule following of fire departments in a Canadian market that was fiercely loyal to Canadian manufacturers. The resort community of Whistler in British Columbia purchased this 95-foot platform on a 1990 Hurricane chassis with unique paint scheme. S/O 6829 was equipped with a 1250 gpm pump and 200-gallon water tank. *Dan Goyer*

In 1992, E-One introduced the HP (High Performance) model platforms which included the HP105, a 105-foot version with a 1000-pound dry tip load and 500-pound rating while flowing up to 2000 gallons per minute. It was stabilized by four H-style outriggers and two jacks that dropped down from just under the cab. Clark County, Nevada, received the pilot model on a Hurricane chassis. S/O 9387 featured a 1500 gpm pump and 150-gallon water tank. *Chuck Madderom*

The 50-foot Teleboom had effectively replaced the older 55-foot StratoSpear. The beachside town of Hatteras, North Carolina, specified the Teleboom on a 1992 International 4900 4x4 chassis in consideration of the sandy environment the apparatus could be called upon to traverse. S/O 10987 was equipped with a 1250 gpm pump and 500-gallon water tank. *Mike Sanders*

By 1991, NFPA had required that all aerial ladders have a minimum 250-pound tip load and E-One upgraded the 110-foot aerial ladder accordingly. Stratford, Ontario, in Canada took delivery of this unit on a 1992 Hurricane chassis. S/O 9344 featured a 1250 gpm pump and 200-gallon water tank. D*ave Stewardson*

The entry level Protector and Protector TC had lighter chassis and power trains, but could accept a 50-foot Teleboom aerial device. Portland, Tennessee, received this rear-mounted unit in 1992 on a Protector TC chassis. S/O 9627 carried 500 gallons of water and was equipped with a 1250 gpm Godiva pump. *Kent Parrish*

The Hush chassis was modified to include a mid-engine configuration in order to accept full size aerial devices. This view shows the standard officer-side engine compartment with grill. Indiana Head, Maryland, received S/O 13008 in 1993 with an HP 95 platform. *Jeff Mogush*

In 1989, E-One redesigned the 80-foot aerial ladder to become more in line with the industry standard of 75 feet. With Superior now under the Federal Signal umbrella, an E-One custom chassis and aerial could be sent to Canada, where Superior would complete the apparatus for its Canadian customer, as in this delivery for Banff, Alberta, on a 1993 Cyclone TC chassis with 1500 gpm pump and 300-gallon water tank. *Shane MacKichan*

When the Protector became E-One's first tilt-cab chassis, it was also beefed up enough to accept a 75-foot aerial ladder if requested by the customer. The lighter 75-foot models could also accommodate a 500-gallon water tank, all on a single rear axle chassis. Cumberland, Rhode Island, received S/O 12952 in 1993 on a Protector TC chassis with a 1250 gpm pump and 500-gallon water tank. *Shaun P. Ryan*

In 1993, E-One introduced the HM (High Maneuverability) 100-foot aerial ladder. It was basically a 110-foot ladder cut down to 100 feet with the same torque box and jack spread. It featured a 250-pound tip load dry and an optional waterway at the tip of the third section. San Marcos, California, received this HM100 in 1994 on a short cab Cyclone TC with a 1500 gpm pump and 250-gallon water tank. It carried S/O 12885. *Chuck Madderom*

Nashville, Tennessee, took delivery of this HP95 in 1995 on a Hurricane chassis. S/O 12695 was equipped with a 2000 gpm pump and 150-gallon water tank. In 1994, the Hurricane became available as a tilt-cab. The tilt-cab could be differentiated from the fixed cab by the presence of two slots between the cab and front bumper that allowed the cab to tilt, but was otherwise indistinguishable at a quick glance. *Kent Parrish*

In 1994, E-One introduced the HP70. It was a 70-foot aerial ladder with two one-person light buckets or "pods," one on each side of the aerial at the tip. The device had a combined tip load of 500 pounds dry and 250 pounds wet. The HP70 did not prove to be popular and was discontinued after approximately ten units were built. Osceola County, Florida, received S/O 1400 in 1994 on a Hurricane chassis with 1500 gpm pump and 500-gallon water tank. *Frank Wegloski*

The E-One Hush was also available for single axle aerials, with either a low profile or standard height cab. The engine remained between the cab and pump with engine access and cooling accomplished on the officer side. Oakland, Maryland, received this 75-foot aerial ladder on a Hush chassis in 1994. It was equipped with a 1500 gpm pump and 500-gallon water tank. *Joel Woods*

Even with a full line of aerial products, E-One continued to offer its customers the choice of a Snorkel product. Lenexa, Kansas, specified this 50-foot Telesqurt on a 1995 Hurricane chassis. S/O 15084 featured a 1250 gpm pump and 500-gallon water tank. *John Arthur*

Delhi Township near Cincinnati, Ohio, received this 50-foot Teleboom on a 1995 Hush chassis. S/O 15095 had a 1250 gpm pump and 500-gallon water tank. The Telesqurt and Teleboom both provided a pumper apparatus with an elevated master stream that served as an alternative to a full size aerial. *Kent Parrish*

In 1995, E-One introduced the CR100 aerial ladder and the HP series a year later. The CR100 was a heavy-duty aerial with an 800-pound tip load. The "CR" acknowledged the crisscross jacks with rectangular extrusions of the original aerial design. The HP ladders were optimized with 500-pound tip loads and in essence replaced the HM100 models. Perry Township near Evansville, Indiana, purchased this demonstrator HP100 in 1996 on a Cyclone TC chassis. S/O 16372 had a 1500 gpm pump and 300-gallon water tank. *Kent Parrish*

This HP105 demonstrator was purchased by St. Louis, Missouri, on a 1996 Hurricane TC chassis. S/O 14359 was equipped with a 1500 gpm pump and 150-gallon water tank. The HP105 would only be available on the Hurricane chassis. *Dennis J. Maag*

The new HP ladder design featured diamond extrusions rather than the original "C" channel or rectangular style. The HP75 was also made available in the American Eagle economy program series. Surprise, Arizona, received this HP75 on an American Eagle custom chassis in 1997. S/O 18217 featured a 1500 gpm pump and 500-gallon water tank. *David Greenberg*

In 1996, Federal Signal purchased the Finnish-built Bronto Sky Lift aerial platform line. The line would fall under several designations and body styles and include heights from 91 to 174 feet. The first several aerials were built in Finland and shipped to E-One, where they were installed on custom chassis. Calgary, Alberta, in Canada purchased this F (Fire Service) HLA (High Level Articulating) 167-foot Bronto on a special 1997 Hurricane six-axle chassis. It was assigned S/O 17410. *Shane MacKichan*

The island of Aruba received this F HDA (Heavy Duty Articulating) 91-foot Bronto in 1998 on a Cyclone II chassis. S/O 17854 had a 1000 gpm Waterous pump and 400-gallon water tank. Earlier E-One Bronto Sky Lifts featured somewhat of a European body configuration. *Shaun P. Ryan*

Up till this point, E-One had not entered the tractor-drawn aerial market, although it had built at least two custom tractors for customers that pulled LTI aerials. In 1998, E-One introduced its first single-source unit. It utilized an HP100 aerial ladder with two jacks mounted at the base of the turntable behind the tractor. Kansas City, Missouri, received the first five units in 1999 on Cyclone II chassis. *Eric Hansen*

West Milford, New Jersey, purchased this HP105 in 1999 on a Hurricane TC chassis. S/O 20433 was equipped with a 1500 gpm pump and 150-gallon water tank. *Neal Van Deusen*

Houston, Texas, was a loyal E-One aerial customer during the mid to late 1990s, purchasing numerous units. The typical ladder specification was a CR100 on a Hurricane TC chassis. Ladder 31 was assigned S/O 121616, a 2000 model. The Hurricane was rarely specified on new pumpers, but its low profile was still favored by many customers for aerials. *Eric Hansen*

Despite fewer Hush model pumpers coming off the line, the mid-engine variation was still commonly chosen for aerial apparatus. Orchard Beach, Maryland, received this HP95 in 2000. *Jeff Mogush*

Penn Township in York County, Pennsylvania, purchased this F HDA 118-foot Bronto on a 2001 Cyclone II chassis. It is equipped with a 2000 gpm pump and 300-gallon water tank. *Juergen Kiefer*

The fire brigade at the Jack Daniels distillery in Lynchburg, Tennessee, operates two E-One apparatus in the company's scheme, including this modified industrial aerial. The bucket of an HP95 was removed, resulting in a heavy-duty 90-foot aerial ladder, and replaced by a 3000 gpm foam nozzle. S/O 123349 is equipped with a 3000 gpm pump and Williams Hot Shot foam system. It carries 500 gallons of foam and 500 pounds of Purple K that can also be applied through a deck gun on each side. *Eric Hansen*

E-One had not offered its full size aerial on a proprietary chassis since the mid 1980s until it was made available in 2000 in the American Eagle economy series on the Spartan Advantage chassis. Del City, Oklahoma, received this HP75 "super-commercial" in 2001 with a 1250 gpm pump and 500-gallon water tank. *Steve Loftin*

Baltimore County Truck 18 was assigned this HP100 tractor-drawn aerial in 2002 on a Cyclone II chassis. It carried S/O 125509. To date, E-One has produced around two dozen tractor-drawn aerials. *Joel Woods*

In 2000, E-One introduced the Side Stacker hose bed for the HP75, which has since been adopted throughout the aerial line. It allowed for storage of 1000 feet of 5-inch supply hose on the officer side and the ability to lay and reload without raising or moving the aerial. St. Matthews, Kentucky, received S/O 125081 in 2002 on a Cyclone II chassis with 1500 gpm pump and 500-gallon water tank. Intakes and discharges were located on the officer side and at the rear. *Kent Parrish*

E-One also introduced the HP95 mid-mount platform in 2000. It was a four-section aerial with 1000-pound tip load. It was available on the Cyclone II chassis and available with a pump, tank, and Side Stacker hose bed. Steelton, Pennsylvania, received this unit in 2003 with a 2000 gpm pump and 300-gallon water tank. *Patrick Rudisil*

The new mid-range Typhoon chassis was a perfect fit to accommodate a 50-foot Teleboom or HP75 aerial. McCulloch, Indiana, received this unit in 2003 with a 1250 gpm pump and 750-gallon water tank. It was assigned S/O 126476. *Kent Parrish*

In 2000, E-One began to refine the body of the Bronto aerials more in accordance with the American fire service. Among the improvements was a redesigned boom to reduce travel height and body reconfigured with ground ladders stored in the aerial torque box. Now a part of the FRG group and noted for its high quality stainless steel bodies, Saulsbury built the body for the new RLP (Rescue Ladder Platform) Bronto available in 100-, 114-, and 134-foot heights. Red Deer, Alberta, the Canadian home of Superior received S/O 126978 on a 2003 Cyclone II chassis with 1750 gpm pump, 240-gallon water tank, and 100-foot Bronto. *Dave Stewardson*

The venerable 110-foot aerial ladder was still available, but generally not the industry standard. The advantage of the 110-foot aerial was a more maneuverable single rear axle chassis. It was still preferred though by larger, traditional departments. Cincinnati, Ohio, exclusively purchased E-One aerials from the early 1990s into the 2000s. Ladder 19 received the first in 1994 and replaced it with this 2004 model. *Joel Woods*

Reno, Nevada, received two of these HP100 tractor-drawn aerials in 2004 on Cyclone II chassis. They were assigned S/O numbers 127776-77. E-One tractor-drawn aerials featured the exclusive V Max tiller cab. The aerial was now rated at 550 pounds wet or dry. *Shane MacKichan*

This mid-mount platform was delivered to Ramsey, New Jersey, in 2004 on a Cyclone II chassis. Without a pump or tank and with a lower travel height, the HP95 mid-mount was designated an HP94. It was assigned S/O 126617. *Neal Van Deusen*

The HP75 aerial ladder was available on the Typhoon custom chassis and in the new Tradition economy program. West McCracken, Kentucky, received this unit in 2004. S/O 129088 was equipped with a 1250 gpm pump and 500-gallon water tank. *Kent Parrish*

Aurora, Colorado, purchased this RLP 100-foot Bronto in 2005 on a Cyclone II chassis. S/O 129295 was not equipped with a pump or tank. The boom was painted red to match the apparatus and carried a Stokes basket mounted to the side. *Stefan Farage*

The Monarch Fire Company in Amity Township, Pennsylvania, operates this striking 2006 Cyclone II with an HP75 aerial ladder. It is equipped with a 2000 gpm pump and 500-gallon water tank. With its lightweight, but strong aluminum design with unique construction and innovative stabilization, E-One is a true pioneer in the modern aerial genre. New products include an HP100 aerial platform to replace the HP105 and the availability of the versatile Snozzle elevated master stream. *Patrick Rudisil*

Chapter 4

The Reedy Creek Improvement District, which was created to maintain and protect the fantasyland of Walt Disney World in Florida, was among Emergency One's earliest repeat customers. In 1976, Reedy Creek took delivery of S/O 185 on a Ford F-350 chassis. The first Emergency One rescue units were basically modified ambulances with diamond plate bodies that had a camper-like appearance. *Garry Kadzielawski*

RESCUE & SPECIALTY APPARATUS

Emergency One was founded on the premise of building ambulances, but soon shifted gears towards fire trucks. However, the popular television show after which the company was partly named had seemingly jump-started the fire service's acceptance of EMS and, in turn, specialty functions. So, it only seemed natural to maintain a presence in the rescue and specialty market. Within just a few years of its inception, the company had introduced one of the most diverse lines in the industry and to date E-One has built rescue and specialty apparatus on practically every chassis imaginable while drifting in and out of the ambulance business.

Early Emergency One rescue models generally correlated to the same chassis configurations associated with the quick attack lineup and amounted to just as many configurations. Oak Forest, Illinois, received this 20084 medium-duty rescue on a 1976 GMC chassis. It carried S/O 285. *Garry Kadzielawski*

The South Dixie Fire District in Louisville, Kentucky, received this 10060 light-duty rescue on a 1977 Ford F-350 chassis with four-door crew-cab. It was assigned S/O 409. *Greg Stapleton*

The state of Illinois was a pioneer in early rescue and combination apparatus. Hoffman Estates took delivery of this 20084TC medium-duty rescue. S/O 619 was built on a 1978 Ford C chassis. *Garry Kadzielawki*

The Casey County Rescue Squad in Kentucky operates this 20084 medium-duty rescue on a 1976 GMC chassis. S/O 823 was painted lime green, a trendy color during the late 1970s that E-One promoted and was thought to possibly make emergency vehicles more visible to motorists. *Kent Parrish*

Ocean Beach, New York, operated a couple of early specialty units built by Emergency One, including this unique hose wagon on a 1979 Chevrolet 4x4 chassis with unpainted body and diamond plate compartment doors. Note the preconnected hard suction running up over the cab into the hose bed via the front-mounted pump. *John Toomey*

The light-duty 10060 was Emergency One's smallest and most popular rescue apparatus. The Belltown Fire Company of Stamford, Connecticut, operated this unit on a 1979 Chevy 4x4 chassis. It carried S/O 1224. *Frank Wegloski*

The 10060 light-duty rescue was also available as a walk-in unit. Prairie-Dupont, Illinois, received S/O 1911 on a 1980 Ford F-350 chassis. Note how the upper portion of the box with windows was set. *Dennis J. Maag*

Skokie, Illinois, received this one-off fire department on wheels on a 1981 Ford C chassis. Featuring a 750 gpm pump and 500-gallon water tank, S/O 1871 also nearly carried enough ground ladders to qualify as a quad. *Garry Kadzielawski*

Mission, Kansas, received this stubby rescue in 1982 on a Ford C chassis, which was a special variation on the 20084TC. It featured a 250 gpm pump and 250-gallon water tank along with a large generator and floodlight with high body. *Steve Loftin*

Another variation of the 20102 medium-duty rescue included a walk-in body, such as on this Civil Defense rig for Ansonia, Connecticut. S/O 2722 was built on a 1982 GMC 4x4 chassis. *Mark Redman*

Another one of the impressive rigs at the West Routt Fire District in Hayden, Colorado, was this monstrous heavy rescue on a Peterbilt 6x6 chassis with manual transmission. It served command, rescue, lighting, and support functions. S/O 3219 was equipped with an electric pump with 100-gallon foam tank, 35-kilowatt generator, and four-bottle air cascade system. *Dennis J. Maag*

E-One utilized the numerical model numbers for rescues for just a short period and then began to designate them by their type. Alhambra, Illinois, received this "full width light rescue" with a walk-in box on a 1983 GMC chassis. It was assigned S/O 3246. *Dennis J. Maag*

The medium-duty "squad rescue" was based on a 21,000 GVW chassis. S/O 3782 was delivered to Canyon Lake, California, in 1984 on a GMC chassis. *Chuck Madderom*

When Emergency One introduced its first municipal custom chassis in 1984, it was immediately adapted to the rescue and specialty line. Widewater, Virginia, received this unique salvage unit in 1985 on a Hurricane chassis. It carried S/O 3442. *Joel Woods*

Emergency Control at the IBM plant in East Fishkill, New York, was accomplished with S/O 4909. The squad rescue was built on a 1986 GMC chassis. The plant also operated a commercial E-One pumper. *Mark Redman*

Flanders, Connecticut, specified this heavy rescue on a 1987 Cyclone custom chassis. S/O 5944 featured a walk-in box and a plethora of compartment space. With the capability to engineer just about anything on wheels, E-One had become a true custom manufacturer and seemingly advertised dozens of variations of rescue and specialty apparatus. *Mark Redman*

Boston was a huge E-One customer, from pumpers to aerials and including specialty apparatus. S/O 5390 was this walk-in extended heavy rescue on a 1987 Ford C chassis assigned to Rescue 2. *Joel Woods*

The Hush rear-engine chassis allowed a perfect environment for the quiet and spacious cab with Vista roof to be utilized as a command post, as it was on this 1987 Hazardous Materials unit for Denver, Colorado. It was assigned S/O 5158. *John Arthur*

Truckee, California, operated an impressive fleet of 4x4 apparatus to protect a mountainous area. S/O 6082, a 1988 walk-in heavy rescue on a Spartan Gladiator 4x4 chassis, was no exception. *Garry Kadzielawski*

Southard, New Jersey, also saw an opportunity to use the advantages of the Hush rear-engine chassis to be utilized as a specialty vehicle as taken on this 1989 delivery that served a multi-purpose function. *Scott Mattson*

Another loyal customer was Phoenix, Arizona, which specified this large Command Unit on a 1989 Duplex Vanguard chassis with bi-fold bus-style crew doors. S/O 8282 featured a separate entrance to the walk-in box with large RV-like windows looking out from a communications center. The entire rig was protected by a full sprinkler system. *David Greenberg*

Salt Lake City, Utah, utilized this medium-duty rig on a 1990 GMC chassis for an air resource/rescue unit. S/O 8358 featured roll-up compartment doors, a feature that would be more common and nearly prevalent in later years. *Stefan Farage*

This tandem axle monster with Vista roof, delivered to New Market, New Jersey, in 1990, was among the largest Hush apparatus ever constructed by E-One. S/O 8016 served as a command and support vehicle. *Scott Mattson*

The Ford Louisville series had been a reliable heavy-duty commercial chassis for fire apparatus. West Windsor, New Jersey, received this walk-in heavy rescue on a 1991 L-8000 chassis. *Scott Mattson*

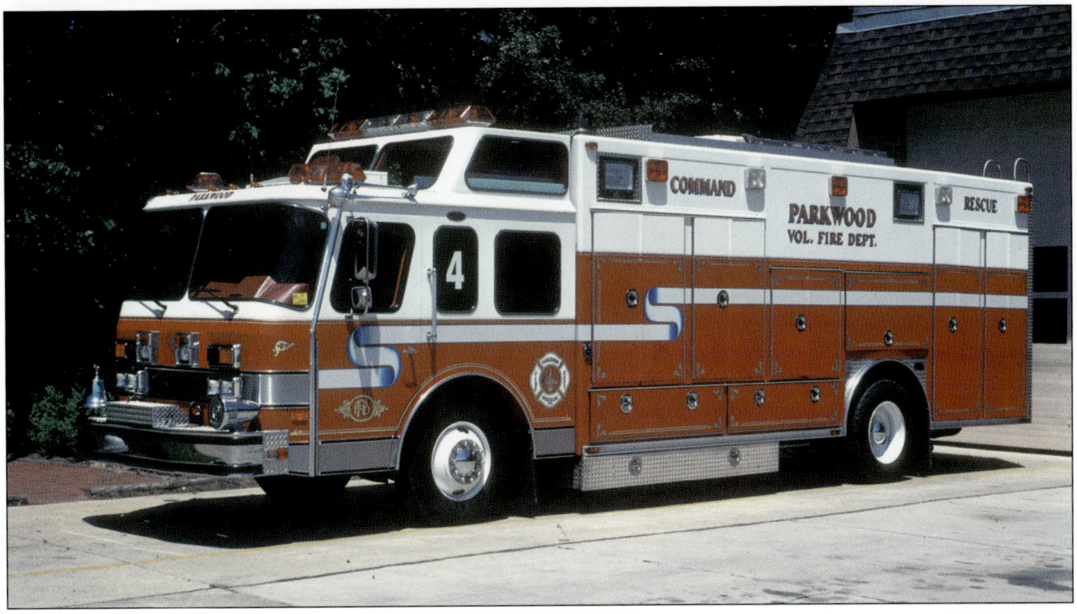

Parkwood, North Carolina, took delivery of this well-equipped command/heavy rescue unit in 1991 on a Hush chassis with Vista roof. S/O 8030 carried a hydraulic generator and light mast in addition to a full complement of rescue tools. *Shaun P. Ryan*

This unique walk-around rescue was built for Camp Jackson, Illinois, on a 1991 Protector chassis. S/O 8746 had the appearance of a pumper with an enclosed pump panel. *Dennis J. Maag*

Colorado departments operated more unique 4x4 or all-wheel-drive E-One apparatus than any state. Lake Dillon specified this walk-in heavy rescue in 1992 on a Cyclone 4x4 chassis. S/O 10746 featured a four-bottle air cascade system. *Dennis J. Maag*

E-One delivered specialty apparatus internationally, in addition to pumpers, aerials, and industrial units. The Hazelwood station in Pretoria, South Africa, received this heavy rescue with walk-in module on a 1992 Hush chassis with Vista roof. It was assigned S/O 10426. *Shaun P. Ryan*

This view shows the rear-engine configuration of a Hush heavy rescue apparatus. S/O 9864 was a 1992 Hush with Vista roof that served as a combination vehicle to provide support, rescue, and haz-mat functions for Schaumburg, Illinois. *Garry Kadzielawski*

Riverside County, California, specified this unique haz-mat unit on a 1993 Cyclone TC chassis with Vista roof. S/O 11009 featured a three-door cab configuration with modified window to accommodate an ergonomic command center and walk-in body. *Chuck Madderom*

Although historically not an E-One customer, Baltimore city purchased this two-toned heavy-duty walk-around unit for Rescue 1 on a 1993 Cyclone TC chassis. *Joel Woods*

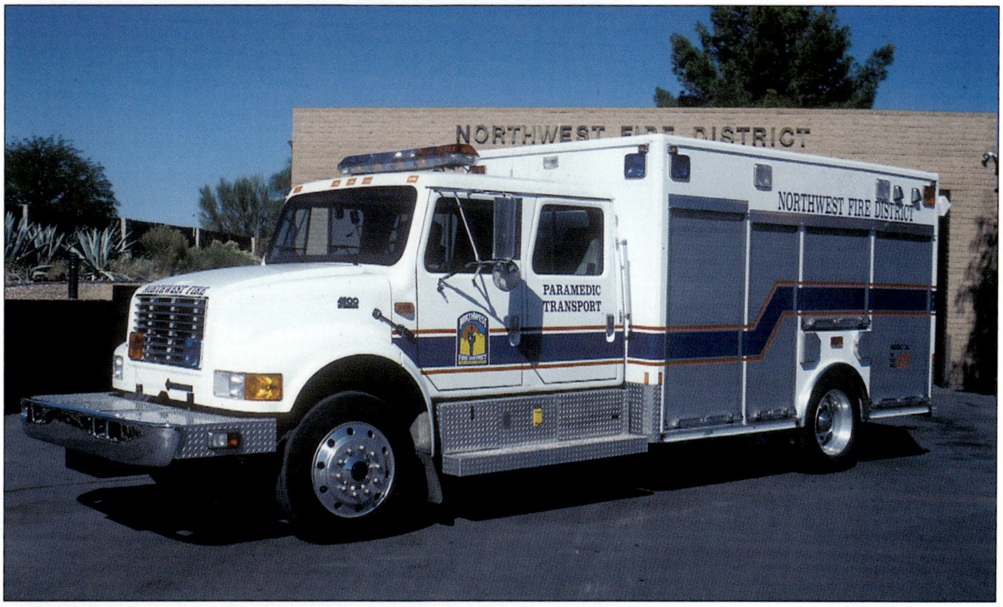

E-One has floated in and out of the ambulance business since 1974. In the 1990s, the EMS/ transport pumper concept was reversed by outfitting a fully functioning ambulance with a firefighting package. The E-One Trident was equipped with a 130 to 250 gpm PTO pump and up to a 300-gallon water tank with optional foam system. The Northwest Fire District in Arizona received this unit in 1994 on an International 4900 chassis. *David Greenberg*

One of the more impressive heavy rescue apparatus built by E-One to date is this monster delivered to Claymont, Delaware, another loyal E-One customer. The tandem axle rig with walk-in body was built on a 1994 Kenworth chassis. S/O 13414 also featured a 750 gpm pump, 350-gallon water tank, and 50-gallon foam cell hidden within the body. *John Toomey*

Gainesville, Florida, specified this haz-mat unit on a 1995 International 4900 chassis with four-door cab. The bulk of the body featured full height and depth compartments for maximum storage and a rear self-contained module that acted as a command/research center. It carried S/O 13982. *Glenn Vincent*

The Protector TC chassis was used for many heavy rescue apparatus as larger horsepower engines were not often necessary and the cost of the specially built body was offset by the choice of a mid-range chassis. S/O 16413 was built for Roosevelt, New York, in 1996. *Neal Van Deusen*

The short-lived entry-level Sentry chassis was used for just a handful of rescue/specialty apparatus, such as this 1996 model for Glendale, Arizona. S/O 16593 was assigned as a Ladder Tender, which relieved the larger aerial apparatus from runs that did not require an aerial and has primarily been a concept in the rapidly growing west. *Garry Kadzielawski*

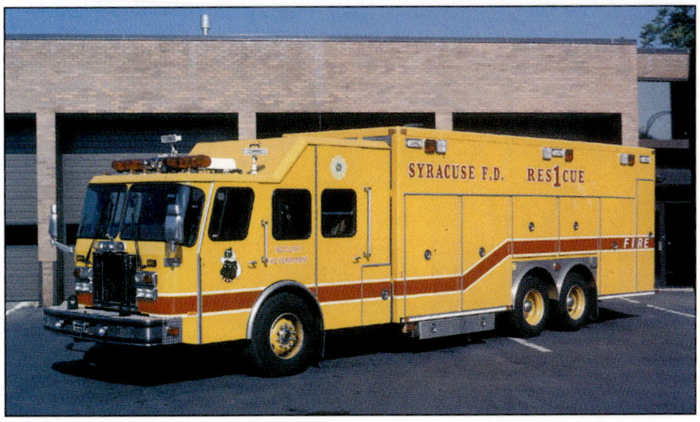

Syracuse, New York, has long operated a dedicated rescue company. Rescue 1 was assigned S/O 17664. This big tandem axle job was built on a 1997 Cyclone II chassis. The Vista raised roof design had now been replaced by a style more integral with the aluminum extrusions of the cab. *Mark Redman*

Clark County, Nevada, operated several Tridents on Freightliner FL-70 chassis. S/O 17184 was a 1998 model. This view shows the 130 gpm PTO pump mounted in the forward portion of the box fed by a 200-gallon water tank and 10-gallon foam cell. The extended front bumper contained a pre-connected attack line. *Chuck Madderom*

This unique combination apparatus was built for Pine-Strawberry, Arizona, on a 1998 International 4700 4x4 chassis. Although assigned as a rescue unit, S/O 18609 featured a 125 gpm pump and 420-gallon water tank with a 60-gallon compressed air foam system (CAFS) that allowed the apparatus to also serve as a quick attack unit. *Garry Kadzielawski*

Norman, Oklahoma, took delivery of this medium-duty rescue with a striking paint scheme on a 1998 International 4900 chassis. It was assigned S/O 19276. Roll-up compartment doors were becoming the norm on rescue and specialty apparatus. *Steve Loftin*

The South Metro Fire District in the Denver area received this tandem axle heavy rescue in 1999 on a Cyclone II chassis. It carried S/O 19838. The amount of specialty equipment being carried by today's rescue teams was resulting in these big apparatus being more common. *John Arthur*

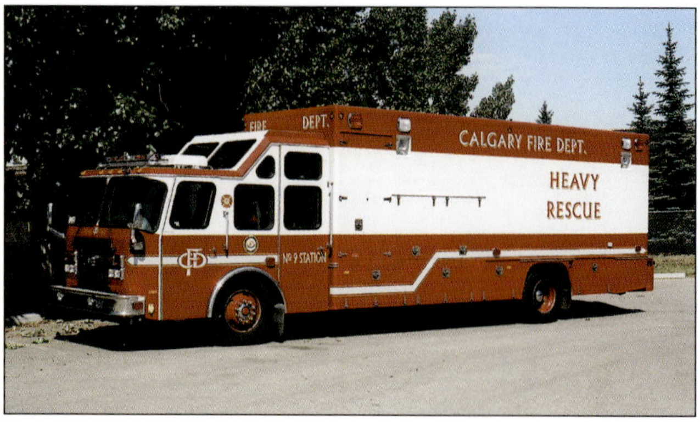

Although the 1992 Federal Signal acquisition of Superior Emergency Vehicles officially linked E-One to the Canadian market, this 1999 Cyclone II chassis was mated with a heavy rescue body built by Interstate Truck and Body for Calgary, Alberta. It carried E-One S/O 21243. *Shane MacKichan*

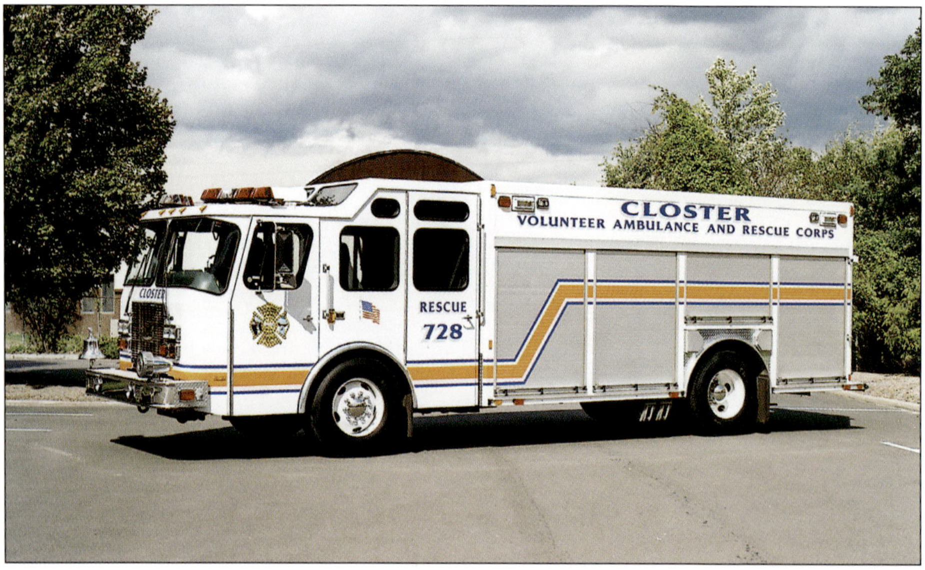

In certain parts of the country, some entities provide heavy rescue functions in addition to serving as an ambulance service. The Closter, New York, Volunteer Ambulance and Rescue Corps purchased this 2000 Cyclone II walk-around rescue. It was assigned S/O 22017. *Stefan Farage*

From 1997 to its closing, the Superior plant in Canada was producing the majority of smaller apparatus under the E-One name, including light duty rescues such as S/O 123172 for Whitesville, Kentucky, built on a 2000 Ford F-550 4x4 chassis with walk-in box. *Kent Parrish*

The Hurricane was fading into history and near extinction from the E-One custom chassis line. However, Lorton, Virginia, specified this 2000 Hurricane TC "Super Rescue" tandem walk-around unit. S/O 122583 was also equipped with a 250 gpm pump and 250-gallon water tank. *Shaun P. Ryan*

The Marshall Volunteer Rescue Squad in Virginia is another that provides heavy rescue functions in addition to ambulance service with S/O 120253, a striking 2001 Cyclone II walk-around rescue. While uncommon years ago, it was now nearly standard for raised roofs on custom cabs. *Mike Sanders*

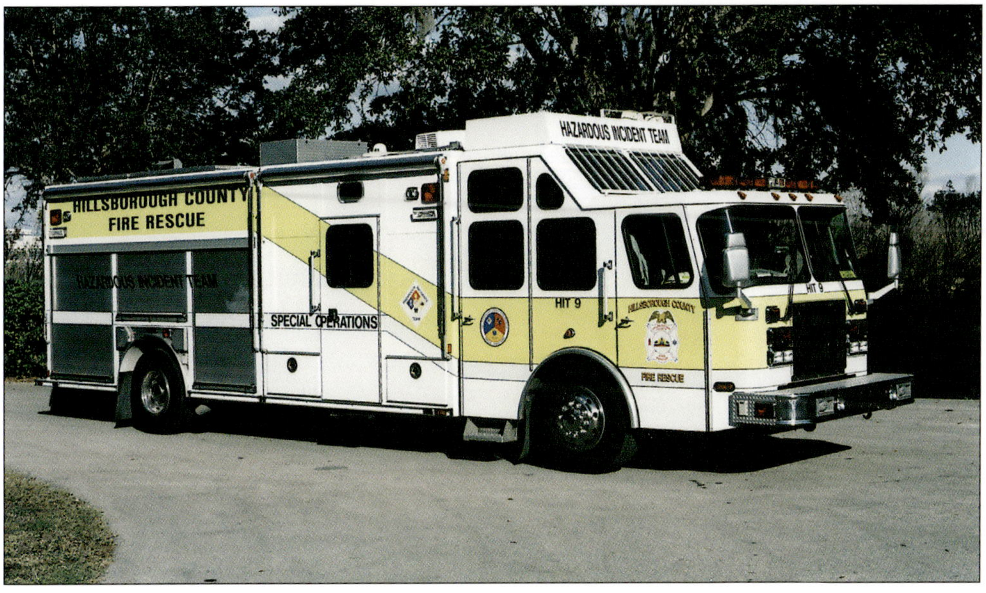

In 1998, Federal Signal acquired Saulsbury Fire Apparatus, which was regarded for not only its use of stainless steel, but as an industry-leading builder of rescue apparatus. Over the next six years, Saulsbury would build apparatus under its own label and on E-One custom chassis such as this aluminum heavy rescue on a 2002 Cyclone II chassis for Hillsborough County, Florida. *Stefan Farage*

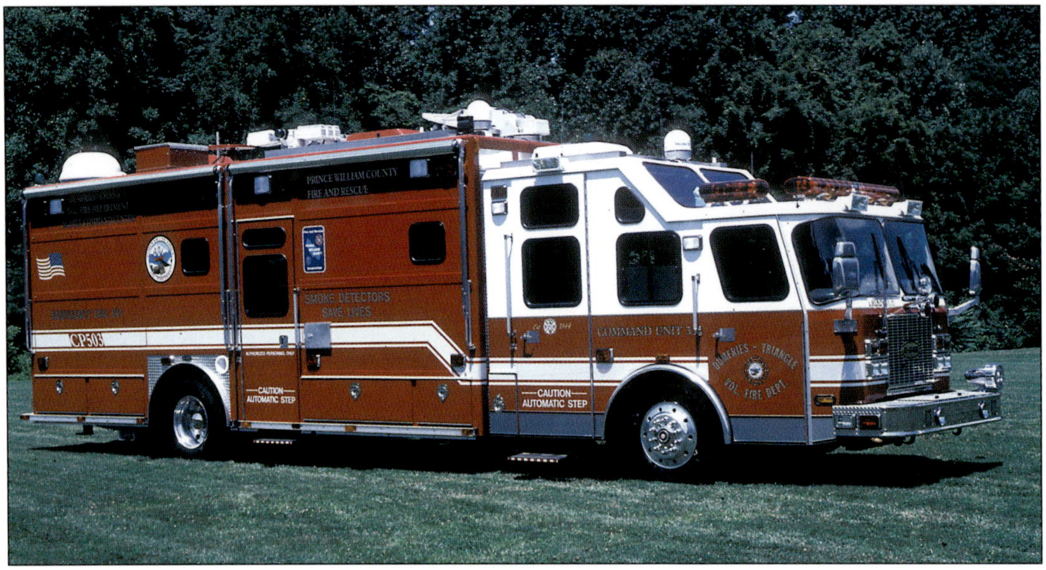

Another volunteer outfit in Virginia, the Dumfries Triangle Volunteer Fire Department, provides command post and support functions with this impressive 2002 Cyclone II. S/O 24467 was in essence an emergency operations center on wheels complete with a fully equipped galley and lavatory facilities. *Joel Woods*

The new E-One Typhoon custom chassis proved to be an immediate success and filled a gap in the mid-range market that E-One had been missing. This aluminum walk-around rescue for Georgetown Township in Floyd County, Indiana, was built by Saulsbury on a 2002 Typhoon chassis. It carried E-One S/O 125194. Federal Signal closed the Saulsbury plant in 2004. *Kent Parrish*

With all the options and electronics of today driving up the cost of a high-end custom chassis, the Typhoon would overtake the flagship Cyclone II as E-One's most popular custom chassis. Marrero-Estelle, Louisiana, received this walk-around heavy rescue on a 2003 Typhoon chassis. *Frank Wegloski*

To date, E-One has produced specialty apparatus for nearly every type of agency imaginable – from riot control vehicles to animal control units. Given the sensitivity of today's Homeland Security, more and more units are being funded by the government. Bomb and fire investigation functions in Montgomery County, Maryland, are provided by S/O 127458 on a 2004 International 4400 chassis. *Joel Woods*

The O.W.L. volunteer fire department in Woodbridge, Virginia, has long been a loyal and noted E-One customer. This monster of a tandem axle heavy rescue was no exception. S/O 129806 was built by Saulsbury out of stainless steel on a 2005 Hurricane TC chassis with modified two-door cab. *Mike Sanders*

S/O 131148 was a light-duty rescue built on a Ford F-550 4x4 chassis at the Superior plant for Langley Township in British Columbia, Canada. By the end of 2006, Federal Signal had closed the Superior plant. *Shane MacKichan*

Swansea, Massachusetts, took delivery of this walk-in heavy rescue on a 2007 Cyclone II chassis. As Federal Signal debates the future of E-One, apparatus continues to roll out of Ocala at a steadier pace with the quality and innovation the company was once known for. No matter its future, E-One has left its mark as a pioneer in not only modern pumpers and aerials, but rescue and specialty vehicles as well. *Dick Barlett*

Chapter 5

In 1982, Emergency One began development of its own CFR apparatus. A prototype was constructed on a British Scammell 4x4 amphibious chassis with an aluminum cab and body. Powered by a Detroit Diesel 8V92T engine, it was equipped with a 1250 gpm pump and 1000-gallon water tank with an "around the pump" foam proportioning system. This CFR was the first to be advertised by Emergency One in literature. The fate of this prototype is unknown. *Byron Brown collection*

AIRCRAFT RESCUE & FIRE FIGHTING RIGS

Aviation firefighting and rescue is handled by apparatus with specifications and operating features put forth by the National Fire Protection Association and Federal Aviation Administration that set them apart from their structural counterparts. During its early years, Emergency One offered variations of its quick attack units that could serve as small CFR (Crash Fire Rescue) units. In 1982, it began development of its own custom ARFF (Aircraft Rescue & Fire Fighting) vehicle, which became known as the Titan series, and began marketing RIV (Rapid Intervention Vehicle) models. During the 1990s, E-One also collaborated under subcontract to produce P-23 ARFF vehicles specifically for the military. To date, E-One ARFF apparatus have been delivered to airports and airfields all over the world. Domestic units continue to roll out of the Florida plant and models with international flair are produced at E-One Europe's plant.

Early CFR apparatus built by Emergency One were variations of the quick attack units on commercial chassis with foam and dry chemical systems. In 1983, Emergency One began marketing a line of RIV units. The RIV-100 was the smallest and featured a utility type body with single or dual agent skid units. The Republic Airport in Farmingdale, New York, operated this 1983 Chevy with a 250 gpm pump, 260-gallon water tank, 20-gallon foam cell, and 100-pound dry chemical system. *Mark Redman*

Emergency One refined the Scammell prototype and introduced its first custom CFR apparatus in 1983 as the Titan. It was available on 4x4 or 6x6 all-wheel-drive custom chassis with ¼-inch aluminum cab and modular body. Water tanks ranged from 1000 to 3000 gallons. Pumps could be driven by a power divider, PTO or separate pump engine. Roof turrets and bumper turrets were available as well as ground sweeps. Anchorage, Alaska, received one of the first deliveries, this 1983 model with a 1250 gpm pump, 1585-gallon water tank, and 190-gallon foam cell. *Mark Redman*

The Washington Dulles International Airport received S/O 3535 in 1984 with a 1250 gpm pump. The Titan III featured a standard 1500-gallon water tank and 200-gallon foam tank. It was engineered with a lower center of gravity for better weight distribution, quick response, and ease of operation by one person. The Titan was technically Emergency One's first exclusive custom chassis as it debuted before the municipal Hurricane, which came a year later. *Mark Redman*

The Mercer County Airport in New Jersey took delivery of the first Emergency One Titan V in 1985. The Titan V variation featured pumps up to 1800 gpm and water tanks up to 3000 gallons on 6x6 custom chassis with 300- to 400-gallon foam tanks. A Deutz Diesel engine was optional in the Titan series. This apparatus was assigned S/O 3482. *Scott Mattson*

Emergency One combined its CFR apparatus with an aerial device to provide another dimension to airport protection. The Greater Buffalo International Airport in New York received this Titan IV Aerial in 1986 with an 1800 gpm pump, 2500-gallon water tank, 250-gallon foam tank, and 1000-pound Halon system. It was equipped with a 55-foot StratoSpear telescoping waterway. With "FAA yellow" the preferred CFR color, red was an unusual choice. *Mark Redman*

The Tucson Airport in Arizona received this Titan III in 1986. S/O 4855 featured a 1250 gpm pump, 1500-gallon water tank, and carried 205 gallons of foam. The Titan series was engineered to accommodate a number of special extinguishing agents, from water and foam to dry chemicals or clean agents necessary for the special hazards posed by aircraft and airports. *Don Feipel*

The RIV-200 was a variation of the 10060 mini-pumper. The body featured a sloped rear for off-road encroachment. It was available with various pump, water tank, and special agent systems in addition to optional turrets. The Wendover Airport in Utah utilized S/O 6285 with a 350 gpm pump, roof turret, 250-gallon water tank, and 30-gallon foam tank. It was assigned S/O 6285. *Shaun P. Ryan*

The McDonnell-Douglas Corporation purchased this RIV-200 for its aircraft construction facility in St. Louis, Missouri. S/O 7406 was built on a 1989 Ford Super-Duty 4x4 chassis. It featured a 250 gpm pump, 250-gallon water tank, 20-gallon foam tank, and 450-pound dry chemical system with a roof mounted turret and pre-connected hose reel. *Dennis J. Maag*

The medium-duty RIV-500 and RIV-1000 were available on 28,000 to 39,000 GVW commercial chassis with 500- to 1000-gallon water tanks. Pumps were 250 to 500 gpm PTO driven or 500 to 750 gpm separate engine driven. The Greater Rochester International Airport in New York received this RIV-500 on a 1989 International S model chassis with a 250 gpm PTO pump, 500-gallon water tank, 60-gallon foam cell, and 500-pound Halitron system. *Mark Redman*

The state of Alaska purchased this Titan III in 1989 for the Wiley Post/Will Rogers Memorial Airport in Barrow. S/O 6945 featured a 1500 gpm pump, 1500-gallon water tank, 190-gallon foam tank, and 700-pound dry chemical system. *Garry Kadzielawski*

The RIV-100 made a nice little unit for smaller airports such as the Barkley Regional Airport in Paducah, Kentucky. S/O 7446 was built on a 1990 GMC 4x4 chassis. It was equipped with a 100-gallon premix of water/foam and a 400-pound Purple K system. *Kent Parrish*

Amarillo International Airport in Texas received this Titan III in 1990. It was equipped with a 1250 gpm pump, 1500-gallon water tank, 190-gallon foam tank, and 500-pound dry chemical system. Higher visibility colors were still preferred on airport apparatus. *David Greenberg*

In 1988, the U.S. Department of Defense awarded a contract to Teledyne Continental Motors to produce 200 "P-23" ARFF vehicles for air bases around the world. Based on an 8x8 all-wheel-drive chassis with Timoney independent suspension system, E-One was subcontracted to the build the aluminum cab and body along with the firefighting system for the P-23. The standard would include a 1500 gpm pump, 3300-gallon water tank, 500-gallon foam tank, and 550-pound Halon system. This unit served as a prototype/demonstrator model. *Joel Woods*

With airports all over the world, E-One's CRF models propelled an already solid international business. The Turkey Ministry of Defense purchased twelve of these blue Titan V 3000-gallon units in 1991. *Joel Woods*

The Yampa Valley Regional Airport in Colorado received two of these Titan III ARFF units in 1992. They were equipped with 1250 gpm pumps, 750-gallon water tanks, 20-gallon foam cells, and 450-pound Purple K systems. The "ARFF" acronym had generally replaced the "CFR" tag. *Dennis J. Maag*

In 1992, the FAA Technical Center in Pomona, New Jersey, received this E-One HPR (High Performance Rescue) 4x4 apparatus for testing purposes and to serve as the center's ARFF apparatus. Based on the independent suspension system adopted from the P-23 military ARFF, E-One had refined its Titan III models from the ground up to be quicker and more efficient. It also included technological advances such as thermal imagers, video cameras, and global positioning. The prototype featured a 1250 gpm pump, 750-gallon water tank, and dual 60-gallon foam tanks with a 500-pound Purple K system. *Mark Redman*

The Myrtle Beach Jetport in South Carolina received this well equipped RIV-200 in 1993 on a GMC 4x4 chassis. The 500 gpm unit featured a dry chemical system in addition to water and foam tank. It also carried a full complement of ground ladders. *Joel Woods*

This RIV-500 was purchased by the DuPage Airport Authority in Illinois and manned by the West Chicago Fire Protection District. S/O 11008 was equipped with a 500 gpm pump, 500-gallon water tank, 60-gallon foam tank, and 300-pound dry chemical system. It was built on a 1993 International 4800 chassis. *Garry Kadzielawski*

E-One had refined the HPR ARFF concept and brought all 4x4 models under the Titan HPR designation. The Albany County Airport in New York received this stubby unit in 1994. It was equipped with a 1250 gpm pump, 750-gallon water tank, 130-gallon foam tank, and 500-pound Purple K system. *Mark Redman*

The Titan IV Aerial had been redesigned in conjunction with the new E-One Teleboom 50-foot telescoping waterway. The device provided a means to reach fires in the high number 2 engines of taller aircraft and was also available in a 35-foot version. Syracuse Hancock International Airport in New York received this unit in 1994 with a 1500 gpm pump, 2500-gallon water tank, and 320-gallon foam tank with a 50-foot Teleboom. Just a handful of the Titan IV Aerials were actually sold. *Mark Redman*

A new model was the RIV-300 combined agent vehicle on a Ford F-Super Duty 4x4 chassis. The new body design was equipped with a 400 gpm rear-mount pump, 300-gallon water tank, 40-gallon foam cell, and 450-pound dry chemical system with a bumper-mounted turret. The Owensboro-Daviess County Regional Airport in Kentucky received this 1996 model. *Greg Stapleton*

The E-One RIV-500 had also been redesigned with a new body style that resembled more of a full size ARFF apparatus than the older quick attack style. McCook Municipal Airport in Nebraska is protected by this red 1996 model on an International 4800 chassis. S/O 16673 is equipped with a 500 gpm pump, 500-gallon water tank, 65-gallon foam cell, and 450-pound Purple K system. *Dennis J. Maag*

After construction and several delays during the stringent U.S. Air Force's acceptance testing and procedures, the P-23 ARFF began deployment to air bases around the world in 1994. Andrews Air Force Base in Maryland received this unit in 1996. Built by E-One on the Teledyne chassis, the P-23 was equipped with features that included a central tire inflation system and in-cab SCBA connections for four riding positions. Approximately 287 of these units were scheduled to be produced under this contract. *Joel Woods*

Many of the principles of the military P-23 were adapted for a civilian version as E-One introduced a larger HPR ARFF apparatus available on 6x6 or 8x8 Titan chassis. Included was an exclusive independent suspension from Timoney Technologies, which was now an E-One subsidiary. Vancouver International Airport in Canada received three Titan HPR 8x8 apparatus in 1996 equipped with 2000 gpm pumps, 3170-gallon water tanks, 405-gallon foam tanks, and 500-pound dry chemical systems. *Dan Goyer*

With three variations in the HRP ARFF line, gone were the numerical designations. E-One simply differentiated the models by their chassis. Victoria International Airport in British Columbia, Canada, purchased this Titan HPR 4x4 in 1996 with a 1050 gpm (Imperial) pump, 1250-gallon water tank, 167-gallon foam tank, and 500-pound dry chemical system. *Dave Stewardson*

The rapidly expanding Dallas-Forth Worth International Airport purchased three E-One Titan HPR 8x8 apparatus in 1996. S/O numbers 16328–30 were equipped with 1000 gpm pumps, 3000-gallon water tanks, 360-gallon foam tanks, and 500-pound Purple K systems. *Frank Wegloski*

The Clark County Fire Department protects visitors to Las Vegas, Nevada, via McCarran International Airport with this 1997 Titan HPR 6x6 with 2000 gpm pump, 3000-gallon water tank, 390-gallon foam tank, and 500-pound Purple K system. S/O 17294 was also equipped with a 50-foot Snozzle from Crash Rescue, which was now available on E-One ARFF apparatus. *Chuck Madderom*

Aberdeen Regional Airport in South Dakota operates this Titan HPR 4x4 with a 1250 gpm pump, 1500-gallon water tank, and 200-gallon foam tank. The compact unit carried S/O 18618. *Larry Phillips*

This Titan HPR 8x8 with 50-foot Snozzle was purchased for the Ontario International Airport in California. S/O 18001 featured a 2000 gpm pump, 3000-gallon water tank, 390-gallon foam tank, and 550-pound Purple K system. *Juergen Kiefer*

Louisville International Airport in Kentucky operates this 1999 Titan HPR 4x4. S/O 19777 features a 1950 gpm pump, 1500-gallon water tank, and 200-gallon foam tank. The 50-foot Snozzle was available with a nozzle that could pierce the skin of an aircraft and act as a distributor as well as thermal imaging unit. *Kent Parrish*

This Titan HPR 6x6 was delivered to the Lehigh Valley International Airport in Pennsylvania. It was equipped with a 2000 gpm pump, 3000-gallon water tank, and 400-gallon water tank. *Mark Redman*

Some fire departments across the country have purchased ARFF apparatus to serve as specialty units. Fairfax County, Virginia, received this white over red Titan HPR 4x4 in 1999 to serve as a foam unit. S/O 20406 was equipped with a 1500 gpm pump, 1500-gallon water tank, 300-gallon foam tank, and 300-pound dry chemical system. *Mike Sanders*

As the P-23 military contract was completed, many branches continued to receive E-One Titan HPR apparatus. The US Army's Fort Campbell on the Kentucky/Tennessee line received S/O 20232 in 1999. The 8x8 model was equipped with a 1500 gpm pump, 3000-gallon water tank, and 400-gallon foam tank. The military was also beginning to switch to a red paint scheme on most new apparatus. *Greg Stapleton*

The U.S. Coast Guard Station in Sacramento, California, received this Titan HPR 4x4 in 2000. It featured a 1500 gpm pump, 1500-gallon water tank, 200-gallon foam tank, and 500-pound dry chemical system. *Shane MacKichan*

Los Angeles International (LAX) Airport in California received this Titan HPR 8x8 in 2000. S/O 120070 was equipped with a 2000 gpm pump, 4000-gallon water tank, 510-gallon foam tank, and 700-pound dry chemical system. It also featured a 50-foot Snozzle. The big Detroit Diesel engine provided 1000 horsepower to move the apparatus. *Chuck Madderom*

This Titan HPR 4x4 was delivered to Renner Field in Goodland, Kansas, in 2001. S/O 121269 was equipped with a 1500 gpm pump, 1000-gallon water tank, 130-gallon foam tank, and 500-pound dry chemical system. *Dennis J. Maag*

By 2003, E-One had redesigned the cab and body of the HPR. The Bellingham International Airport in Washington received this 4x4 in 2003 with a 1250 gpm pump, 1585-gallon water tank, 202-gallon foam tank, and 500-pound dry chemical system. The new Titan appeared a bit boxier than the older design and featured a more perpendicular cab windshield and just driver and passenger door windows. *Shane MacKichan*

The new E-One HPR "G" series is produced by E-One Europe, a subsidiary in Stellendam, the Netherlands. The European influenced model consists of a cab, body, and tanks built entirely of glassfiber-reinforced polyester (GRP). This 2004 HPR-G was delivered to the Volkel Air Force Base in Uden, the Netherlands, with a 2100 gpm pump, 3055-gallon water tank, 198-gallon foam tank, and 550-pound dry chemical system. The Netherlands ordered a total of 35 units. *Laurens Bontenbal*

With Phoenix, Arizona, being such a loyal E-One customer, it was no exception that Sky Harbor International Airport utilized an E-One ARFF fleet including this red 2005 Titan HPR 8x8 model. S/O 128216 featured a 2000 gpm pump, 3200-gallon water tank, 400-gallon foam tank, and 700-pound dry chemical system. *David Greenberg*

The Waterloo International Airport in Iowa purchased this Titan HPR 4x4 in 2006. S/O 130647 was equipped with a 1500 gpm pump, 1500-gallon water tank, 205-gallon foam tank, and 450-pound dry chemical system with a Hydro-Chem roof turret. *Don Feipel*

This Titan HPR 8x8 was delivered to the Kansas City, Missouri, International Airport in 2006. S/O 129674 featured a 2000 gpm pump, 3170-gallon water tank, 400-gallon foam tank, and 500-pound Purple K system. It was equipped with a 50-foot Snozzle. *Dennis J. Maag*

The Piedmont-Greensboro International Airport in North Carolina received this Titan HPR 4x4 in 2007 with a 1500 gpm pump, 1500-gallon water tank, 205-gallon water tank, and 500-pound Purple K system. Since 1983, E-One has provided a specialized apparatus for airfields and airports all over the world and continues to do so with the Titan line, including the European-built HPR-G and a new Ocala-built Force F6 model. *Shaun P. Ryan*

EMERGENCY VEHICLES

100 Years of American LaFrance: An Illustrated History ISBN 1-58388-139-5
Ahrens-Fox Fire Apparatus Photo Archive ISBN 1-58388-217-0
American Fire Apparatus Co. 1922-1993 Photo Archive ISBN 1-58388-131-X
American LaFrance 400 Series Photo Archive ISBN 1-58388-185-9
American LaFrance 500/600 Series: Photo Archive ISBN 1-58388-210-3
American LaFrance 700 Series 1945-1952 Photo Archive ISBN 1-882256-90-5
American LaFrance 700 Series 1945-1952 Photo Archive Volume 2 ISBN 1-58388-025-9
American LaFrance 700 & 800 Series 1953-1958 Photo Archive ISBN 1-882256-91-3
American LaFrance 900 Series 1958-1964 Photo Archive ISBN 1-58388-002-X
Classic Seagrave 1935-1951 Photo Archive ISBN 1-58388-034-8
Crown Firecoach 1951-1985 Photo Archive ISBN 1-58388-047-X
Emergency One Fire Apparatus 1974-2007 ISBN 1-58388-225-1
Elevating Platforms: A Fire Apparatus Photo Gallery ISBN 1-58388-164-6
Encyclopedia of Canadian Fire Apparatus ISBN 1-58388-119-0
Fire Chief Cars 1900-1997 Photo Album ISBN 1-882256-87-5
Firefighting Tanker Trucks and Tenders: A Fire Apparatus Photo Gallery ISBN 1-58388-138-7
Funeral Service & Flower Cars Photo Archive ISBN 1-58388-227-8
FWD Fire Trucks 1914-1963 Photo Archive ISBN 1-58388-156-5
Grumman Fire Apparatus 1976-1992 Photo Archive ISBN 1-58388-165-4
Hahn Fire Apparatus 1923-1990 Photo Archive ISBN 1-58388-077-1
Heavy Rescue Trucks 1931-2000 Photo Gallery ISBN 1-58388-045-3
Imperial Fire Apparatus 1969-1976 Photo Archive ISBN 1-58388-091-7
Industrial and Private Fire Apparatus 1925-2001 Photo Archive ISBN 1-58388-049-6
*Mack Fire Trucks 1911-2005 Illustrated History** ISBN 1-58388-157-3
Maxim Fire Apparatus 1914-1989 Photo Archive ISBN 1-58388-050-X
Maxim Fire Apparatus Photo History ISBN 1-58388-111-5
Navy & Marine Corps Fire Apparatus 1836 -2000 Photo Gallery ISBN 1-58388-031-3
Pierre Thibault Ltd. Fire Apparatus 1918-1990 Photo Archive ISBN 1-58388-074-7
Pierce Fire Apparatus 1939-2006 Illustrated History ISBN 1-58388-189-1
Pirsch Fire Apparatus 1890-1991 Photo Archive ISBN 1-58388-082-8
Police Cars: Restoring, Collecting & Showing America's Finest Sedans ISBN 1-58388-046-1
Saulsbury Fire Rescue Apparatus 1956-2003 Photo Archive ISBN 1-58388-106-9
Seagrave 70th Anniversary Series Photo Archive ISBN 1-58388-001-1
Seagrave Fire Apparatus 1959-2004 Photo Archive ISBN 1-58388-132-8
TASC Fire Apparatus 1946-1985 Photo Archive ISBN 1-58388-065-8
The Illustrated Encyclopedia of American Fire Engine Manufacturers ISBN 1-58388-203-0
Van Pelt Fire Apparatus 1925-1987 Photo Archive ISBN 1-58388-143-3
Volunteer & Rural Fire Apparatus Photo Gallery ISBN 1-58388-005-4
W.S. Darley & Co. Fire Apparatus 1908-2000 Photo Archive ISBN 1-58388-061-5
Wildland Fire Apparatus 1940-2001 Photo Gallery ISBN 1-58388-056-9
Young Fire Equipment 1932-1991 Photo Archive ISBN 1-58388-015-1

AUTOMOTIVE

AMC Cars 1954-1987: An Illustrated History ISBN 1-58388-112-3
AMC Performance Cars 1951-1983 Photo Archive ISBN 1-58388-127-1
AMX Photo Archive: From Concept to Reality ISBN 1-58388-062-3
The American Taxi- A Century of Service ISBN 1-58388-176-X
Avanti: The Complete Story ISBN 1-58388-213-8
Buick 1946-1960 Photo Archive ISBN 1-58388-178-6
Cadillac: The Tailfin Years ISBN 1-58388-212-X
Camaro 1967-2000 Photo Archive ISBN 1-58388-032-1
Checker Cab Co. Photo History ISBN 1-58388-100-X
Chevrolet Corvair Photo History ISBN 1-58388-118-2
Chevrolet Station Wagons 1946-1966 Photo Archive ISBN 1-58388-069-0
Classic American Limousines 1955-2000 Photo Archive ISBN 1-58388-041-0
Cobra and Shelby Mustang Photo Archive 1962-2007 Including Prototypes and Clones ISBN 1-58388-198-0
The Complete U.S. Automotive Sales Literature Checklist 1946-2000 ISBN 1-58388-155-7
Corvair by Chevrolet Experimental & Production Cars 1957-1969, Ludvigsen Library Series ISBN 1-58388-058-5
Corvette The Exotic Experimental Cars, Ludvigsen Library Series ISBN 1-58388-017-8
Corvette Prototypes & Show Cars Photo Album ISBN 1-882256-77-8
DeTomaso Pantera ISBN 1-58388-177-8
Duesenberg Racecars and Passenger Cars Photo Archive ISBN 1-58388-145-X
Encyclopedia of Small-Scale Diecast Motor Vehicle Manufacturers ISBN 1-58388-174-3
Ferrari- The Factory Maranello's Secrets 1950-1975, Ludvigsen Library Series ISBN 1-58388-085-2
Ford Postwar Flatheads 1946-1953 Photo Archive ISBN 1-58388-080-1
Ford Station Wagons 1929-1991 Photo History ISBN 1-58388-103-4
Grand Prix: Pontiac's Luxury Performance Car ISBN 1-58388-184-0
Hudson Automobiles 1934-1957 Photo Archive ISBN 1-58388-110-7
Imperial 1964-1968 Photo Archive ISBN 1-882256-23-9
It's Delightful! It's Delovely! It's DeSoto Automobiles ISBN 1-58388-172-7
Jaguar XK120, XK140, XK150 Sports Cars Ludvigsen Library Series ISBN 1-58388-150-6
Javelin Photo Archive: From Concept to Reality ISBN 1-58388-071-2
The Lincoln Continental Story: From Zephyr to Mark II ISBN 1-58388-154-9
Lincoln Motor Cars 1920-1942 Photo Archive ISBN 1-882256-57-3
Mercedes-Benz 300SL: Gullwings and Roadsters 1954-1964 Ludvigsen Library Series ISBN 1-58388-137-9
Mercury Automobiles 1939-1959 ISBN 1-58388-205-7
MG Saloons and Coupes 1925-1980 ISBN 1-58388-144-1
Nash 1936-1957 Photo Archive ISBN 1-58388-086-0
Oldsmobile 1946-1960 Photo Archive ISBN 1-58388-168-9
Packard Motor Cars 1946-1958 Photo Archive ISBN 1-882256-45-X
Pontiac Dream Cars, Show Cars & Prototypes 1928-1998 Photo Album ISBN 1-882256-93-X
Pontiac's Greatest Decade 1959-1969—The Wide Track Era: An Illustrated History ISBN 1-58388-163-8
Rambler 1950-1969 Photo Archive ISBN 1-58388-078-X
Stretch Limousines 1928-2001 Photo Archive ISBN 1-58388-070-4
Studebaker Lark 1959-1966 Photo Archive ISBN 1-58388-107-7
The Collector's Guide to GTO 1964-1974 ISBN 1-58388-196-4
Thirty Years of the Volkswagen Golf & Rabbit ISBN 1-58388-158-1
Ultimate Corvette Trivia Challenge ISBN 1-58388-035-6

BUSES

Buses of ACF Photo Archive Including ACF-Brill And CCF-Brill ISBN 1-58388-101-8
Buses of Motor Coach Industries 1932-2000 Photo Archive ISBN 1-58388-039-9
City Transit Buses of the 20th Century Photo Gallery ISBN 1-58388-146-8
Fageol & Twin Coach Buses 1922-1956 Photo Archive ISBN 1-58388-075-5
Flxible Intercity Buses 1924-1970 Photo Archive ISBN 1-58388-108-5
Flxible Transit Buses 1953-1995 Photo Archive ISBN 1-58388-053-4
GM Intercity Coaches 1944-1980 Photo Archive ISBN 1-58388-099-2
Greyhound in Postcards: Buses, Depots and Posthouses ISBN 1-58388-130-1
Highway Buses of the 20th Century Photo Gallery ISBN 1-58388-121-2
*Mack® Buses 1900-1960 Photo Archive** ISBN 1-58388-020-8
New York City Transit Buses 1945-1975 Photo Archive ISBN 1-58388-149-2
Prevost Buses 1924-2002 Photo Archive ISBN 1-58388-083-6
Rapid Transit Series Buses: General Motors and Beyond ISBN 1-58388-209-X
Trailways Buses 1936-2001 Photo Archive ISBN 1-58388-029-1
Trolley Buses 1913-2001 Photo Archive ISBN 1-58388-057-7
Trolley Buses Around the World: A Photo Gallery ISBN 1-58388-175-1
Welcome Aboard the GM New Look Bus: An Enthusiast's Reference ISBN 1-58388-167-5
Yellow Coach Buses 1923-1943 Photo Archive ISBN 1-58388-054-2

RAILWAYS

Chicago & North Western Passenger Trains of the 400 Fleet Photo Archive ISBN 1-58388-159-X
Chicago Stations & Trains Photo Archive ISBN 1-58388-216-2
Chicago, St. Paul, Minneapolis & Omaha Railway 1880-1940 Photo Archive ISBN 1-882256-67-0
Chicagoland Commuter Railroads: Metra & Northern Indiana Commuter Transportation District ISBN 1-58388-190-5
Classic Sreamliners Photo Archive: The Trains and the Designers ISBN 1-58388-144-x
Freight Trains of the Upper Mississippi River Photo Archive ISBN 1-58388-136-0
Great Lakes Ore Docks and Ore Cars ISBN 1-58388-202-2
Great Northern Railway Ore Docks of Lake Superior Photo Archive ISBN 1-58388-073-9
Illinois Central Railroad 1854-1960 Photo Archive ISBN 1-58388-063-1
Interurban Trains to Chicago Photo Archive ISBN 1-58388-199-9
Locomotives of the Upper Midwest Photo Archive: Diesel Power in the 1960s and 1970s ISBN 1-58388-113-1
Milwaukee Road 1850-1960 Photo Archive ISBN 1-882256-61-1
Milwaukee Road Depots 1856-1954 Photo Archive ISBN 1-58388-040-2
Northern Pacific Railway Photo Archive ISBN 1-58388-186-7
Pacific Coast Commuter Railroads: From San Diego to Anchorage ISBN 1-58388-221-9
Pennsylvania Railroad Locomotives: Steam, Diesel & Electric Photo Archive ISBN 1-58388-228-6
Show Trains of the 20th Century ISBN 1-58388-030-5
Soo Line 1975-1992 Photo Archive ISBN 1-882256-68-9
Steam Locomotives of the B&O Railroad Photo Archive ISBN 1-58388-095-X
Trains of the Circus 1872-1956 ISBN 1-58388-024-0
Trains of the Upper Midwest Photo Archive Steam & Diesel in the 1950s & 1960s ISBN 1-58388-036-4

RECREATIONAL VEHICLES & OTHER

Commercial Ships on the Great Lakes Photo Gallery ISBN 1-58388-153-0
Drag Boats of the 1960s Photo Archive ISBN 1-58388-222-7
Illustrated Guide to Snowmobile Racing ISBN 1-58388-170-0
Phillips 66 1945-1954 Photo Archive ISBN 1-882256-42-5
RV & Camper Toys: The History of RVing in Miniature ISBN 1-58388-207-3
RVs & Campers 1900-2000: An Illustrated History ISBN 1-58388-064-X
Ski-Doo Racing Sleds 1960-2003 Photo Archive ISBN 1-58388-105-0
The Collector's Guide to Ski-Doo Snowmobiles ISBN 1-58388-133-6
Tugboats of the East Coast: A Photo Gallery ISBN 1-58388-224-3
Tugboats of the Great Lakes ISBN 1-58388-192-1
Vintage Snowmobila: A Guide to Snowmobile Collectibles ISBN 1-58388-169-7
Vintage Snowmobiles: The Golden Years 1968-1982 ISBN 1-58388-193-X

More Great Titles From **Iconografix**

All Iconografix books are available from direct mail specialty book dealers and bookstores worldwide, or can be ordered from the publisher. For book trade and distribution information or to add your name to our mailing list and receive a **FREE CATALOG** contact:

Iconografix, Inc.
PO Box 446, Dept BK, Hudson, WI, 54016

Telephone: (715) 381-9755, (800) 289-3504 (USA), Fax: (715) 381-9756
info@iconografixinc.com
www.iconografixinc.com